Optics

Problems and Solutions

Antonio Siciliano

University of Bari, Italy

Illustrations by
Francesca Perniola

Optics

Problems and Solutions

World Scientific

NEW JERSEY · LONDON · SINGAPORE · BEIJING · SHANGHAI · HONG KONG · TAIPEI · CHENNAI

Published by

World Scientific Publishing Co. Pte. Ltd.

5 Toh Tuck Link, Singapore 596224

USA office: 27 Warren Street, Suite 401-402, Hackensack, NJ 07601

UK office: 57 Shelton Street, Covent Garden, London WC2H 9HE

British Library Cataloguing-in-Publication Data
A catalogue record for this book is available from the British Library.

OPTICS
Problems and Solutions

ISBN-13 978-981-256-796-3
ISBN-10 981-256-796-8
ISBN-13 978-981-256-842-7 (pbk)
ISBN-10 981-256-842-5 (pbk)

Printed in Singapore

to
Antonio
Carlo
Bianca
Beatrice

Preface

This book concerns light as a *moving energy*: the main feature of a *light source* is its power (energy emitted per unit time); the prominent and distinctive characteristic of light (if *quasi-monochromatic*), observed compulsorily on a surface, is its intensity (energy reaching a unit surface per unit time) and for *natural* light is its spectral intensity (energy hitting a unit surface per unit time and per unit frequency). Throughout each chapter these are the dominant and recurring arguments. The emphasis on light (see Appendix 1) as moving energy grants continuity, either it must be considered a wave or the *moving entity* called photon. In addition deliberately no mention has been made to old and new optical technologies. These references seem misleading for students because technologies are a very serious matter (reserved to engineering universities and treated in different textbooks) and cannot be deceptively used as attracting argument to study and understand the basic concepts of Optics.

In the mid-1950s, as a student, and in the first years of 1960s, as a professor, we used a slide rule (an unknown instrument to the present-day students) to do simple calculations. Then cumbersome and enormous calculators (mainframes and minicomputers) appeared and were used as a little more sophisticated tool of calculus and for simpler tasks pocket calculators. Later in 1980s the Personal Computers begin to appear and in 1990s Internet. These are now standard and *universal utilities*. Nevertheless textbooks propose scientific problems for students as if the only available calculus instrument is a pocket calculator.

In this book MATLAB (a trademak of MathWorks) is used. Someone can rightly use another equivalent program: it is a choice the author has to do. But MATLAB or an equivalent program *is a must*. Without this tool the book wouldn't appear as it is. In many circumstances it has been an unyielding and implacable corrector of some substantial or material mistakes or oversights.

With this tool finding roots of equations, solving differential equations, determining the value of an integral with integrand function of real or complex arguments, mastering matrix calculus, plotting and other current subjects of numerical mathematics become a real affordable task for a student.

As example, relevant to our book, consider the *reference to untraceable tables* to find values of the Bessel (1784-1846) integrals used by Airy (1801-1892) to define diffraction for a circular aperture or values of Fresnel (1788-1827) integrals used by Cornu (1841-1902) to plot in a complex plane a spiral. Both arguments seem *nebulous* to students. These eminent scientists excellently used the available tools of their times. Now we have the chance to do a better choice.

In the book appear *only the results* of a long list of MATLAB scripts. Their presence in the book would double the number of pages and confuse, with necessary links, the linear sequence of the content. However readers can refer to scripts using a complementary virtual booklet (www.optics-as.com) that will also include corrections, to unavoidable errors present in this book, supplementary problems, and readers' suggestions.

Usually a list of problems is included at the end of each chapter of a Physics textbook and answers are given to some of them. This position is an indicator of their ancillary condition versus academic lessons. In science the experimental observation is the starting point for theory; so in the educational process the problems would introduce professorial lectures. The book is an attempt to give prior focus to problems in order to seize and firmly hold the basic concepts underlying them. Therefore a teacher, using this textbook, can explain and extend the contents of introductions to the main laws present in each chapter (adding some demonstrations, if necessary) meanwhile the problems are discussed and resolved.

Who are we indebted to? Mainly to a long list of students we have encountered in the last forty years and of scientists (their names are the titles of a large number of problems) who, investigating Optics in the last three centuries, have given us a wealth of knowledge about light.

Special thanks must be expressed to Professor Mauro Castellani for his invaluable help to accomplish this project.

Contents

Chapter 1: Reflection and Refraction

Chapter 2: Lenses

2.1 Main Laws and Formulae

2.2. Problems

Chapter 5: Diffraction

Chapter 6: Photons and Moving Light Sources

Chapter 1

Reflection and Refraction

1.1 Main Laws and Formulae

1.1.1 The laws of reflection and refraction

a. The directions of the incident ray R_1, the reflected ray R'_1 and the normal N to the surface AB (Fig. 1.1) are coplanar. The angle of incidence α_1

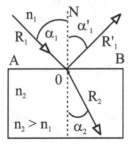

Fig. 1.1

and the angle of reflection α'_1 are equal.

b. The directions of the incident ray R_1, the refracted ray R_2 and the normal N (Fig 1.1) are coplanar. The angle of incidence α_1 and the angle of refraction α_2 are related by the law

$$n_1 \sin \alpha_1 = n_2 \sin \alpha_2 \qquad (1.1)$$

where n_1 and n_2 are called refractive indices of the two transparent, uniform and isotropic media separated by the plane surface AB.

If the medium above the surface AB is denser than that below it ($n_1 > n_2$), the angle of refraction α_2 will always be greater than the angle of incidence α_1. So there is a limit angle $\alpha_{1\,max}$ that allows refraction given by

$$\alpha_{1\,max} = \arcsin\left(\frac{n_2}{n_1}\right) \tag{1.2}$$

For $\alpha_1 > \alpha_{1\,max}$ there will be only reflection.

1.1.2 Sign convention
a. Assuming a Cartesian coordinate system a light ray moves from left to right along the z axis in the plane yz (Fig. 1.2). The height of an object is positive if it is above the plane xz.

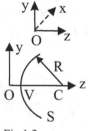

Fig.1.2

b. If S (Fig. 1.2) is a section in the plane yz of a spherical surface of center C and radius R the right line through the points V and C is assumed as z axis and called optic axis. More spherical surfaces having the same optic axis form a coaxial system.

c. The radius R is positive/negative if S presents its convex/concave surface to the incident ray.

d. We consider only rays which make small angles with z axis. These are called paraxial. With this assumption we can write

$$\sin\alpha = \alpha \quad \tan\alpha = \alpha \quad \cos\alpha = 1 \tag{1.3}$$

d. The angle between a ray and z axis is positive if its trigonometric tangent is positive, lengths being measured from the corner C where the right angle is situated (Fig. 1.3).

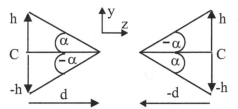

Fig. 1.3 Positive/negative values of h, d and α

1.1.3 Reflection and refraction on a spherical surface

The laws of reflection and refraction relate angles. Their application to spherical surfaces establishes a reciprocal relationship between distances z of an object and z' of an image from a spherical surface (Fig. 1.4). Two corresponding points of object and image are called conjugate.

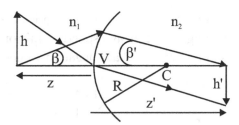

Fig. 1.4

For a refracting spherical surface z, z', h, h', β and β' are related by the following relations

$$\frac{n_2}{z'} = \frac{n_1}{z} + \frac{n_2 - n}{R}$$
(1.4.1)

$$M_T = \frac{h'}{h} = \frac{n_1 z'}{n_2 z}$$
(1.4.2)

$$M_\beta = \frac{\beta'}{\beta} = \frac{z}{z'}$$
(1.4.3)

where M_T and M_β are called transverse and angular magnification. For a reflecting spherical surface (concave/convex mirrors) the previous relations become

$$\frac{1}{z'}=-\frac{1}{z}+\frac{2}{R} \tag{1.5.1}$$

$$M_T=-\frac{z'}{z} \tag{1.5.2}$$

$$M_\beta=-\frac{1}{M_T}=\frac{z}{z'} \tag{1.5.3}$$

For z approaching infinite from the (1.5.1) we have the focal length of spherical mirrors

$$f=\frac{R}{2} \tag{1.6}$$

where R is negative/positive for concave/convex mirrors.

An example of ray tracing for a concave spherical mirror is given in Fig. 1.5 where $R = OC$ and $f = OF = R/2$ are negative values. The rays $1'$, $2'$

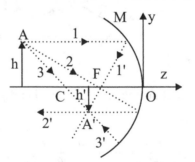

Fig. 1.5 Ray tracing for a spherical concave mirror

and $3'$ are the corresponding reflected rays of the incident rays 1, 2 and 3.

1.1.4 Matrix and vector form for light rays

The matrix treatment of light rays is useful when an optical system, assumed as a serial combination of many simple constituent, must be examined. If the optical elements have the individual matrices

$$[M_1], [M_2], [M_3],..., [M_n] \tag{1.7}$$

the final ray emerging from the optical system is given by the matrix multiplication

$$[M] = [M_n]...[M_3][M_2][M_1] \tag{1.8}$$

We consider now the ray matrices for the simplest optical components.

a. *A ray traveling through an homogeneous medium* (Fig. 1.6) can be set

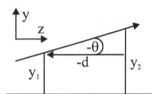

Fig. 1.6 A ray moves in a single homogeneous medium

forth in the two following statements

$$y_2 = y_1 + (-d) \times (-\theta) \tag{1.9.1}$$

$$\theta = \theta_1 = \theta_2 \tag{1.9.2}$$

that can also be easily obtained multiplying the matrix on the right side of (1.10) by the adjacent vector

$$\begin{vmatrix} y_2 \\ \theta \end{vmatrix} = \begin{vmatrix} 1 & d \\ 0 & 1 \end{vmatrix} \begin{vmatrix} y_1 \\ \theta \end{vmatrix} \tag{1.10}$$

b. *A ray reflected by a plane mirror* described by the following statements (Fig. 1.7)

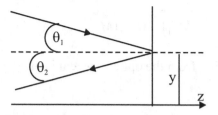

Fig. 1.7 A ray reflected by a plane mirror

$$y = y_1 = y_2 \quad \theta = \theta_2 = -\theta_1 \tag{1.11}$$

can assume the following matrix form

$$\begin{vmatrix} y \\ \theta_2 \end{vmatrix} = \begin{vmatrix} 1 & 0 \\ 0 & 1 \end{vmatrix} \begin{vmatrix} y \\ -\theta_1 \end{vmatrix} \tag{1.12}$$

c. For *a refracted ray from a plane surface*, the statements (Fig. 1.8)

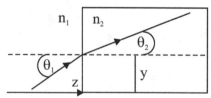

Fig. 1.8 A ray refracted by a plane surface

$$y = y_1 = y_2 \quad \theta_2 = \frac{n_1}{n_2} \theta_1 \tag{1.13}$$

become

$$\begin{vmatrix} y \\ \theta_2 \end{vmatrix} = \begin{vmatrix} 1 & 0 \\ 0 & \dfrac{n_1}{n_2} \end{vmatrix} \begin{vmatrix} y \\ \theta_1 \end{vmatrix} \tag{1.14}$$

d. For *a ray reflected by a spherical surface* the statements (Fig. 1.9)

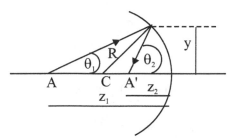

Fig. 1.9 A ray reflected by a spherical surface

$$y = y_1 = y_2 \tag{1.15}$$

$$\frac{1}{z_2} = -\frac{1}{z_1} + \frac{1}{f} \quad \Rightarrow \quad \frac{\theta_2}{y} = -\frac{\theta_1}{y} + \frac{1}{f} \tag{1.16}$$

become

$$\begin{vmatrix} y \\ \theta_2 \end{vmatrix} = \begin{vmatrix} 1 & 0 \\ \dfrac{1}{f} & -1 \end{vmatrix} \begin{vmatrix} y \\ \theta_1 \end{vmatrix} \tag{1.17}$$

e. For *a ray refracted by a spherical surface* (Fig. 1.10) we have

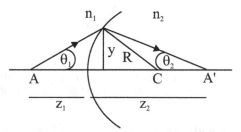

Fig. 1.10 A ray refracted by a spherical surface

$$y = y_1 = y_2 \tag{1.18}$$

$$\frac{n_2}{z_2} = \frac{n_1}{z_1} + \frac{n_2 - n_1}{R} \tag{1.19.1}$$

$$\frac{n_2\,\theta_2}{y} = \frac{n_1\,\theta_1}{y} + \frac{n_2 - n_1}{R} \tag{1.19.2}$$

$$\theta_2 = (1 - \frac{n_1}{n_2})\frac{1}{R}y + \frac{n_1}{n_2}\theta_1 \tag{1.19.3}$$

and the corresponding matrix form is

$$\begin{vmatrix} y \\ \theta_2 \end{vmatrix} = \begin{vmatrix} 1 & 0 \\ (1 - \frac{n_1}{n_2})\frac{1}{R} & \frac{n_1}{n_2} \end{vmatrix} \begin{vmatrix} y \\ \theta_1 \end{vmatrix} \tag{1.20}$$

1.1.5 Reflection and refraction as electromagnetic waves propagation

A light ray is really a propagation of oscillating electric and magnetic fields (Fig. 1.11) with the electric field E perpendicular to the magnetic

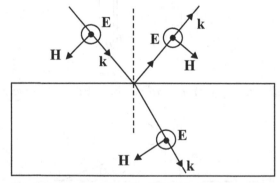

Fig. 1.11 At a fixed time the field H in the plane of the figure and E normal to the plane and directed upwards

field H and both perpendicular to the direction of propagation, defined by the wave vector k having the magnitude $2\pi/\lambda$. A light ray conveys energy that is usually considered per unit time and per unit surface and called intensity or, in the jargon of the Fourier transform, power spectrum or spectral power density. As an example, when a light ray gives rise to reflection and refraction, portion of its intensity is carried by the reflected ray and the remainder by the refracted ray assuming that there is not absorption of energy by the second medium.

The intensity is proportional to the square of the magnitude of the electric field by means of the constant A (see Appendix 1)

$$I = A \cdot E^2 \tag{1.21}$$

Considering a light ray as propagation of electric and magnetic fields, the basic laws of reflection and refraction can be derived from the Fresnel equations under the following assumptions:
a. *the media are homogeneous and isotropic*
b. *the waves are plane and harmonic*
c. *the electric and magnetic fields are continuous* at the boundary surface of the two media.
If the direction of electric field E_1 of the incident ray makes an angle different from 90° with the plane of figure (plane of incidence) the magnitude of its components will be E_{1w} in this plane and E_{1y} normal to this plane. The corresponding magnitudes of the components of the electric field for the reflected (E'_{1w}, E'_{1y}) and refracted (E_{2w}, E_{2y}) rays are given by the following Fresnel equations

$$E'_{1w} = E_{1w} \frac{\tan(\alpha_1 - \alpha_2)}{\tan(\alpha_1 + \alpha_2)} \tag{1.22.1}$$

$$E'_{1y} = -E_{1y} \frac{\sin(\alpha_1 - \alpha_2)}{\sin(\alpha_1 + \alpha_2)} \tag{1.22.2}$$

$$E_{2w} = E_{1w} \frac{2\sin\alpha_2 \cos\alpha_1}{\sin(\alpha_1 + \alpha_2)\cos(\alpha_1 - \alpha_2)} \tag{1.22.3}$$

$$E_{2y} = E_{1y} \frac{2\sin\alpha_2 \cos\alpha_1}{\sin(\alpha_1 + \alpha_2)} \tag{1.22.4}$$

with α_1 the angle of incidence and α_2 the angle of refraction, related by the Snell's law (1.1).

1.1.6 Harmonic waves. Principle of superposition. Complex notation
The simplest wave model having a sinusoidal shape is called harmonic (Fig. 1.12). For a progressive harmonic wave the oscillations along the y

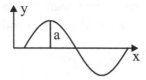

Fig. 1.12

axis, function of x and t, are given by

$$y = a\sin(kx - \omega t) \tag{1.23}$$

where

$$k = \frac{2\pi}{\lambda} \quad \omega = \frac{2\pi}{T} = 2\pi\nu \quad v = \frac{\omega}{k} \tag{1.24}$$

with k the magnitude of the wave vector, λ the wavelength, ω the angular frequency, T the period, ν the frequency and v the velocity of propagation.

The varying values of y in (1.23) represent an electric or a magnetic field. Because the Maxwell's equations for vacuum are linear differential equations, if at a point P in empty space different sources produce the (electric, for example) fields

$$\vec{F}_1, \vec{F}_2, \vec{F}_3, ..., \vec{F}_n \tag{1.25}$$

the resulting field in P is equal to the vector sum

$$\vec{F} = \vec{F}_1 + \vec{F}_2 + \vec{F}_3, ..., + \vec{F}_n \tag{1.26}$$

The optical interference, for exsample, is based on this principle of linear superposition of electromagnetic fields.

Therefore, for example, if at a point P and time t there is a superposition of a progressive wave

$$y_1 = a\sin(kx - \omega t) \tag{1.27}$$

and a regressive one

$$y_2 = a\sin(kx + \omega t) \qquad (1.28)$$

the resultant wave in P at a time t will be

$$y_P = y_1 + y_2 =$$
$$= a\sin(kx - \omega t) + a\sin(kx + \omega t) = \qquad (1.29)$$
$$= 2a\cos\omega t\sin kx$$

and the corresponding intensity at time t will be (A is a constant)

$$I_P = A \cdot 4a^2 \cos^2 \omega t \qquad (1.30)$$

The results (1.29) and (1.30) are usually obtained representing the harmonic wave with complex numbers. Remembering the Euler formula

$$e^{i\varphi} = \cos\varphi + i\sin\varphi$$

we can write

$$y_P = y_1 + y_2 = a(e^{i(kx - \omega t)} + e^{i(kx + \omega t)}) = \qquad (1.31.1)$$
$$= e^{ikx}(e^{-i\omega t} + e^{i\omega t}) =$$

$$= a(\cos kx + i\sin kx)2\cos\omega t = \qquad (1.31.2)$$
$$= 2a\cos kx\cos\omega t + i2a\sin kx\cos\omega t$$

The result (1.29) appears here as the coefficient of the imaginary part of the complex number (1.31.2).

In a straightforward way the intensity is determined using complex notation for the waves. In fact the main purpose in optics in not to ascertain the resulting waveform (1.29) from superposition of the component waves in a point P but the whole energy or intensity (1.30) in this point. Multiplying the complex value (1.31.1)

$$y_P = e^{ikx}(e^{-i\omega t} + e^{i\omega t})$$

by its complex conjugate

$$y^*{}_P = e^{-ikx}(e^{+i\omega t} + e^{-i\omega t})$$

we have directly the intensity (1.30), omitting the constant A)

$$I = y_P y^*{}_P = ae^{ikx}(e^{-i\omega t} + e^{i\omega t})\, ae^{-ikx}(e^{+i\omega t} + e^{-i\omega t})$$

$$a^2(1 + e^{-2i\omega t} + e^{+2i\omega t} + 1) = a^2(1 + 2\cos 2\omega t + 1) =$$

$$= 2a^2(1 + \cos 2\omega t) = 4a^2\cos^2 \omega t$$

The resulting waveform (1.29) represents (Fig. 1.13) a stationary wave.

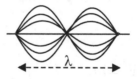

Fig. 1.13

Its profile expands and shrinks vertically but does not move forwards or backwards. In a stationary wave the energy is not transferred in either direction. There is a power density localized in the space but not uniformly and constantly distributed.

1.2 Problems

1.2.1 A simple concave mirror

An object y has a distance z from the vertex O of a concave spherical mirror having focus F and center C (Fig. 1.14). Use ray tracing to find images

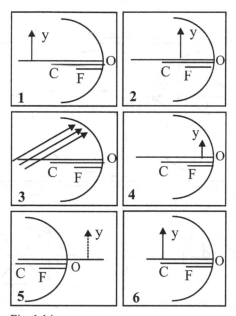

Fig. 1.14

corresponding to the six situations of Fig. 1.14. With the focal distance f = -15 mm determine the image distance z' from O and the magnification m_T varying z, with a step of 5mm, in the interval (-35, 10) mm. Give the plots of z' and m_T varying z.

Solution:

The ray tracing is given in Fig. 1.15. The table 1.1 and Fig. 1.16 give the values of z' and m_T and the corresponding plots.

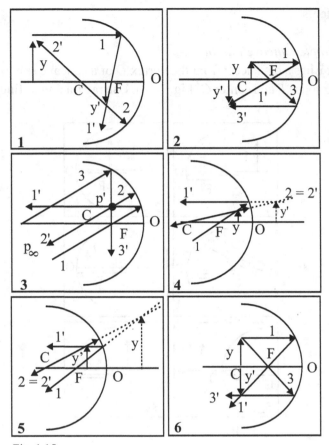

Fig. 1.15

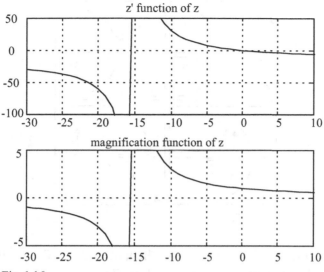

Fig. 1.16

Table 1.1

z (mm)	z' (mm)	m_T
-35	-26.25	-0.75
-30	-30	-1
-25	-37.51	-1.5
-20	-60	-3
-15	∞	∞
-10	30	3
-5	7.5	1.5
0	0	0/0
5	-3.75	0.75
10	-6	0.6

1.2.2 A simple convex mirror

An object y has a distance z from the vertex O of a convex spherical mirror having focus F and center C (Fig. 1.17). Find with ray tracing the images corresponding to the four situations of Fig. 1.17. With the focal length $f = 15$ mm determine the image distance z' from O and the magnification m_T varying z, with a step of 5 mm, in the interval (-30, 30) mm. Give the corresponding plots.

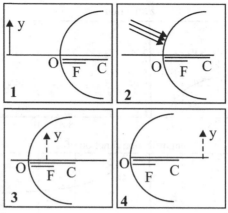

Fig. 1.17.

Solution:

The ray tracing is given in Fig. 1.18. The table 1.2 and Fig. 1.19 contain the values z' and m_T and the corresponding plots.

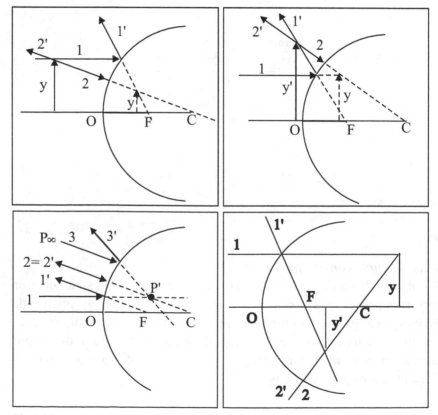

Fig.1.18

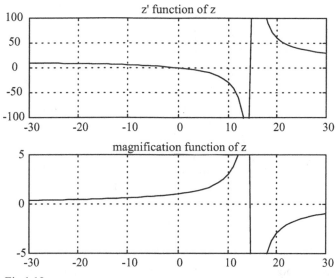

Fig.1.19

Table 1.2

z (mm)	z' (mm)	m_T
-30	10	0.33
-25	9.38	0.38
-20	8.57	0.43
-15	7.50	0.50
-10	6.00	0.60
-5	3.75	0.75
0	0	0/0
5	-7.50	1.50
10	-30	3.00
15	$-\infty$	$-\infty$
20	60.00	-3.00
25	37.50	-1.50
30	30	-1.00

1.2.3 A refracting surface

A box (Fig. 1.20) is filled with water ($n = 4/3$). An object whose height is $y = 6$ m is placed at a distance $z = -10$ m from the point O of the plane

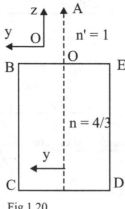

Fig.1.20

boundary between water and air ($n' = 1$).
Determine the position z' and the height y' of the image seen by an observer located in A. Give the corresponding ray tracing.

Solution:

For a refracting spherical surface we have

$$\frac{n'}{z'} = \frac{n}{z} + \frac{n' - n}{R} \qquad M_T = \frac{n z'}{n' z}$$

If the refracting surface is plane (R is infinite) the previous become

$$\frac{n'}{z'} = \frac{n}{z} \qquad z' = \frac{n' z}{n} = \frac{1 \times (-10)}{4/3} = -7.5\,\text{m}$$

$$M_T = \frac{n z'}{n' z} \quad \rightarrow \quad \frac{y'}{y} = \frac{n z'}{n' z} \qquad y' = \frac{n z'}{n' z} y = \frac{n}{n' z} z' y = \frac{1}{z'} z' y = y = 6\,\text{m}$$

The observer perceives the object in the water at a distance $z' = -7.5$ m but having the real height.
Rotating of 90° clockwise the box the ray tracing is simpler (Fig. 1.21).

1.2.4 Prism 1

A ray of monochromatic light is incident at A (Fig. 1.22) with an angle β on a prism whose refractive index is n. Let α be the angle at the edge B.

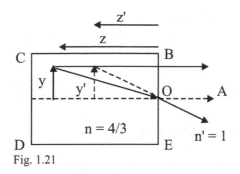

Fig. 1.21

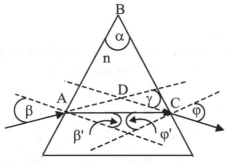

Fig. 1.22

Prove that the angle of deviation between incident beam on A and the emerging beam from C is $\gamma = (n - 1)\alpha$ if α is small.

Solution:
Using the law of refraction in A and in C

$$\sin\beta = n\sin\beta' \tag{1.32}$$

$$\sin\varphi = n\sin\varphi' \tag{1.33}$$

From triangle ABC

$$\beta' + \varphi' = \alpha \tag{1.34}$$

If α is small, it follows from (1.34) that also β' and φ' are small. Hence from (1.32) and (1.33) also β and φ are small. Then

$$\beta = n\beta' \qquad \varphi = n\varphi'$$

From triangle ADC

$$\gamma = (\beta - \beta') + (\varphi - \varphi') = (\beta + \varphi) - (\beta' + \varphi') \qquad (1.35)$$

Using the previous formulae, (1.35) becomes

$$\gamma = n\beta' + n\varphi' - (\beta' + \varphi') = (n-1)(\beta' + \varphi') = (n-1)\alpha$$

1.2.5 A reflecting plate

A parallel beam of monochromatic light is incident with an angle θ_1 on the surface AC (Fig. 1.23) of a box of a transparent and homogeneous

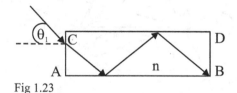

Fig 1.23

material. Find the minimum value of refractive index n which allows total internal reflection on the surfaces AB and CD for every value of θ_1.

Solution:
On the separation surface AC we have

$$\sin \theta_1 = n \sin \theta_2 \qquad (1.36)$$

On lateral surfaces must be (Fig. 1.24)

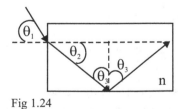

Fig 1.24

$$n\sin\theta_3 = 1 \qquad \theta_3 = \arcsin(\frac{1}{n}) \qquad (1.37)$$

which define the minimum θ_3 that allows total internal reflection. From geometry, (1.36) becomes

$$\sin\theta_1 = n\sin(90° - \theta_3) \qquad \sin\theta_1 = n\cos\theta_3$$

Hence (Fig. 1.25)

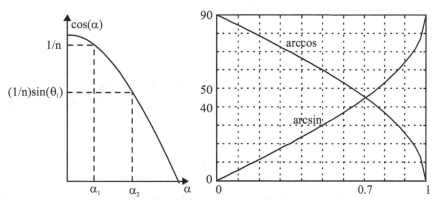

Fig.1.25

$$\theta_3 = \arccos(\frac{1}{n}\sin\theta_1) \geq \arccos(\frac{1}{n}) \qquad (1.38)$$

Using (1.37) and (1.38)

$$\theta_3 = arcsin(\frac{1}{n}) = \arccos(\frac{1}{n}sin\theta_1) \geq \arccos(\frac{1}{n})$$

Because (Fig. 1.25)

$$\arcsin(\frac{1}{n}) \geq \arccos(\frac{1}{n})$$

only if $\sin(\alpha)$ is greater than 0.7071, it follows

$$\frac{1}{n} \geq 0.7071 \qquad\qquad n \leq 1.4142 \qquad (1.39)$$

1.2.6 Prism 2

A ray of monochromatic light is incident at right angle on the face AB
(Fig. 1.26) of a prism having a vertex angle α and a refractive index n.

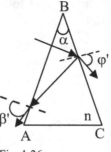

Fig. 1.26

The beam is partially reflected on the face BC and refracted with an angle
$\varphi' = 7.26°$. The reflected ray is refracted again on the face AB with an
angle $\beta' = 14.58°$.
Find values of α and n.

Solution:

The angles φ and α are equal, as both complementary of γ (Fig. 1.27)

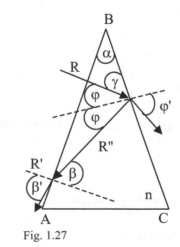

Fig. 1.27

$$\varphi = \alpha \tag{1.40}$$

From geometry

$$\beta = 2\varphi \tag{1.41}$$

From the law of refraction on the faces CB and AB

$$n\sin\varphi = \sin\varphi' \tag{1.42}$$

$$n\sin\beta = \sin\beta' \tag{1.43}$$

Using (1.40) and (1.41), (1.42) and (1.43) become

$$n\sin\alpha = \sin\varphi' \qquad n\sin 2\alpha = \sin\beta'$$

and dividing

$$\frac{n\sin 2\alpha}{n\sin\alpha} = \frac{\sin\beta'}{\sin\varphi'} \qquad 2\cos\alpha = \frac{\sin\beta'}{\sin\varphi'}$$

$$\alpha = \arccos(\frac{\sin\beta'}{2\sin\varphi'}) = \arccos(0.996) = 5.1°$$

Hence from (1.42) using (1.40) we have

$$n = \frac{\sin\varphi'}{\sin\varphi} = \frac{\sin\varphi'}{\sin\alpha} = 1.42$$

1.2.7 A refracting plate
A ray of monochromatic light is incident, with an angle α, on a plate (Fig. 1.28) having a constant thickness $s = 9$ mm and a refractive index

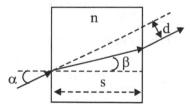

Fig.1.28

$n = 1.56$. Compute for the maximum value of α allowing the transmission of the light through the plate, the corresponding β_{max} and d_{max}. Are β_{max} and/or d_{max} functions of n?

Solution:
On the first face is (Fig. 1.29)

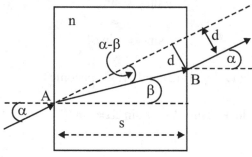

Fig. 1.29

$$\sin\alpha = n\sin\beta \tag{1.44}$$

By geometry

$$s = AB\cos\beta \qquad AB = \frac{s}{\cos\beta} \tag{1.45}$$

and

$$d = AB\sin(\alpha - \beta) \tag{1.46}$$

Substituting from (1.45) AB into (1.46)

$$d = \frac{s}{\cos\beta}\sin(\alpha - \beta) \tag{1.47}$$

For $\alpha = \alpha_{max} = 90°$ (1.47) becomes

$$d_{max} = \frac{s}{\cos\beta}\cos\beta = s = 9\,\text{mm}$$

The value of d_{max} is not dependent from n.
The corresponding maximum value of β (Fig. 1.30) is

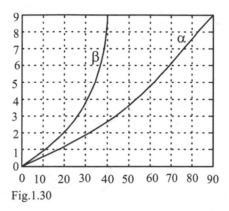

Fig.1.30

$$\beta_{max} = \arcsin(\frac{1}{n}) = 39.9°$$

which is dependent from n.

1.2.8 Prism 3
A ray of monochromatic light is incident (Fig. 1.31) horizontally on a

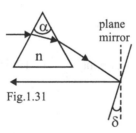

Fig.1.31

prism, whose section is an equilateral triangle, having a refractive index $n = 1.5$. The ray emerging from the prism must be reflected by a plane vertical mirror that must be rotated by an angle δ so that the beam has to return back horizontally. Calculate δ.

Solution:
The plane section of the prism is an equilateral triangle, hence $\alpha = 60°$.
By law of refraction (Fig. 1.32)

$$\sin\beta = n\sin\beta' \qquad\qquad (1.48)$$

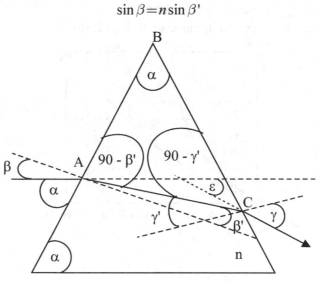

Fig. 1.32

By geometry

$$\beta = 90° - \alpha = 30° \qquad\qquad (1.49)$$

Using (1.48) and (1.49 we have)

$$\beta' = \arcsin(\frac{\sin\beta}{n}) = 19.5°$$

In addition by geometry

$$\alpha = \beta' + \gamma' \qquad\qquad \gamma' = \alpha - \beta' = 40.5°$$

and from the law of refraction

$$n\sin\gamma' = \sin\gamma$$

Hence

$$\gamma = \arcsin(n\sin\gamma') = 77.0°$$

Once more by geometry for the angle of deviation ε

$$\varepsilon = (\beta - \beta') + (\gamma - \gamma') = \beta + \gamma - \alpha = 47.0°$$

From Fig. 1.33 follows

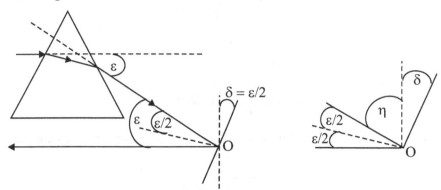

Fig. 1.33

$$\delta = \varepsilon/2 = 23.5°$$

1.2.9 A hemisphere

A spherical concave mirror, whose section is a semicircle of radius $R =$ OC = 30 cm, is filled with water ($n = 1.33$). A ray of monochromatic light is incident, with an angle $\alpha = 4°$, on the superior surface of water. Fixing an xyz system of coordinates with origin in O (Fig. 1.34) determine the position (x, y, z) of the image.

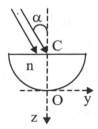

Fig.1.34

Solution:
From the law of refraction (Fig. 1.35) we obtain

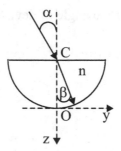

Fig. 1.35

$$\beta=\arcsin(\frac{\sin\alpha}{n})=0.0698=3.01°$$

Fig. 1.36 shows a ray (r_1) crossing the focus F and a ray (r_2) the center C

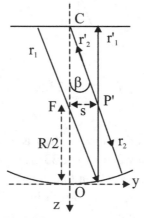

Fig.1.36

of the mirror with corresponding reflected rays $(r'_1$ and $r'_2 = r_2)$.
From geometry follows

$$s=\frac{R}{2}\tan\beta=0.8\,\text{cm}$$

The coordinates of the point image P' are $x = 0$, $y = s = 0.8$cm and $z = -R/2 = -15$ cm

1.2.10 Plane and spherical surfaces
A spherical concave mirror, having radius $R = -30$ cm, is filled with a

solid, homogeneous and transparent material M of refractive index $n' = 1.55$ (Fig. 1.37). Assuming $d = -5$ cm, find the position z' of the image of

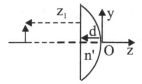

Fig. 1.37

an object having distance, from external and vertical face of M, $z_1 = -100$ cm, the object lateral magnification and the focus of the spherical concave mirror filled with M.

Solution:
The optical system is compounded of a plane refracting surface and a mirror.
At the plane refracting surface we have (Fig. 1.38) for the position z'_1

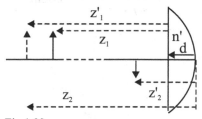

Fig.1.38

$$\frac{n'}{z'_1} = \frac{n}{z_1} \qquad z'_1 = \frac{n'}{n} z_1 = -155 \, \text{cm}$$

Fig. 1.39 gives the related ray tracing. The virtual image y' becomes an object at distance

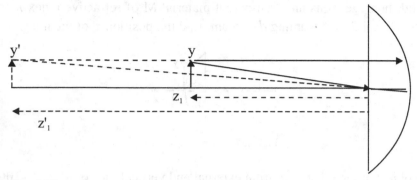

Fig. 1.39

$$z_2 = z'_1 + d = -160\,\text{cm}$$

from the mirror.
The corresponding distance of the image is

$$z'_2 = \frac{f\, z_2}{-f + z_2} = -16.6\,\text{cm}$$

with $f = R/2$.
The magnification is

$$m_T = -\frac{z'_2}{z_2} = -0.1$$

The final real image is reversed (Fig. 1.38) and its height is one tenth of
that of the real object.
The focus of the spherical concave mirror filled with M is

$$f_{\text{new}} = \frac{z\,z'}{z + z'} = -14.2\,\text{cm}$$

assuming $z = z_1 = -100\,\text{cm}$ and $z' = z'_2 = -16.6$ cm.
The focus f_{new} is dependent on d (Fig. 1.40); on the contrary the focus of
a simple spherical mirror

$$f = \frac{R}{2}$$

varies only with *R*.

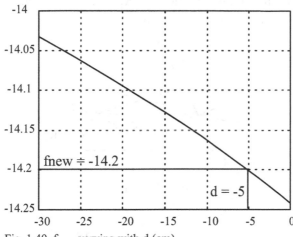

Fig. 1.40 f_{new} varying with d (cm)

1.2.11 Prism 4

A beam of light R_{1-2}, consisting of two wavelengths λ_1 and λ_2, is incident on a prism with apex angle $\alpha = 60°$. The refractive indexes for λ_1 and λ_2 are respectively $n_1 = 1.618$ and $n_2 = 1.652$. Assume that the prism is in position of minimum deviation for the ray R_1 corresponding to the light of wavelength λ_1 (Fig. 1.41). Calculate the angular deviations γ_1 e γ_2 of the rays R_1 and R_2 from the original direction of ray R_{1-2}. Assume that a

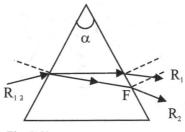

Fig. 1.41

convergent thin lens ($f = 15$ cm) has its optical axis coincident with R_2.and the front focus on point F where the ray R_2 emerges from prism. Find the distance z' from the lens of the point image of R_1 and its height y' from the optical axis.

Solution:

The position of minimum deviation for R_1 (Fig. 1.42) needs

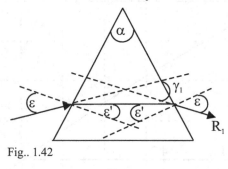

Fig.. 1.42

$$\varepsilon' = \frac{\alpha}{2} = 30°$$

$$\sin \varepsilon = n_1 \sin 30° \qquad \varepsilon = \arcsin(\frac{1.618}{2}) = 54.0°$$

$$\gamma_1 = 2\varepsilon - \alpha = 48.0°$$

For R_2 we have (Fig. 1.43),

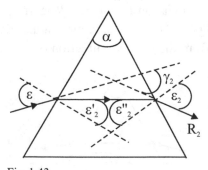

Fig. 1.43

$$\sin \varepsilon = n_2 \sin \varepsilon'_2 \qquad \varepsilon'_2 = \arcsin(\frac{\sin \varepsilon}{n_2}) = 29.3°$$

From the geometry (Fig. 1.43)

$$\varepsilon'_2 + \varepsilon''_2 = \alpha \qquad \varepsilon''_2 = 60 - 29.3 = 30.7°$$

$$n_2 \sin \varepsilon''_2 = \sin \varepsilon_2 \qquad \varepsilon_2 = 57.5°$$

and

$$\gamma_2 = \varepsilon + \varepsilon_2 - \alpha = 54.0 + 57.5 - 60 = 51.5°$$

From the geometry (Fig. 1.44)

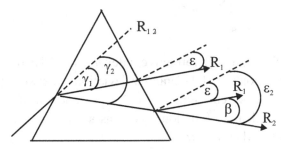

Fig. 1.44

$$\beta = \varepsilon_2 - \varepsilon = 57.5 - 54.0 = 3.5°$$

or

$$\beta = \gamma_2 - \gamma_1 = 51.5 - 48.0 = 3.5°$$

Fig. 1.45 gives $z' = f = 15$ cm and

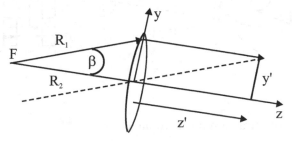

Fig. 1.45

$$y' = 15 \cdot \tan \beta = 0.92 \, cm$$

1.2.12 A sphere

A ray of monochromatic light is incident with angle α (Fig. 1.46) on a

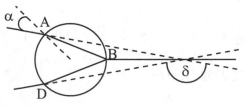

Fig.1.46

transparent sphere, having a refractive index $n = 1.66$. There is on B a partial or total reflection? Find the angular deviation δ, between the ray incident on A and the ray emerging from D, as a function of α. Define the value α_{min} for which the function $\delta(\alpha)$ has a minimum. Determine the values of β and δ when $\alpha = \alpha_{min}$.

Solution:

The angle of incidence on B is equal to the angle of refraction on A as the triangle ACB is isosceles (Fig. 1.47).

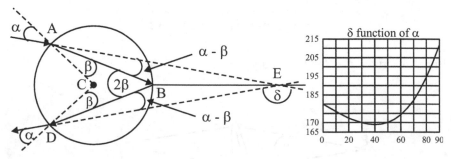

Fig.1.47

From the law of refraction on A we have

$$\beta = \arcsin(\frac{1}{n}\sin\alpha) \qquad (1.50)$$

The angle of total reflection would be

$$\beta *=\arcsin(\frac{1}{n})$$ (1.51)

Because from (1.50) and (1.51)

$$\frac{1}{n}\sin\alpha<\frac{1}{n}$$

and

$$\beta<\beta*$$

it follows that there is only partial reflection on B.
The angular deviations on A, B and C are respectively α -β, π -2β and α -β and their sum is

$$\delta=\pi+2\alpha-4\beta$$ (1.52)

We find the critical value of δ, using (1.52), as a function of α

$$\frac{d\delta}{d\alpha}=\frac{d}{d\alpha}(\pi+2\alpha-4\beta)=$$

$$=2-4\frac{d\beta}{d\alpha}=2-4\frac{d}{d\alpha}(\arcsin(\frac{\sin\alpha}{n}))=$$

$$=2-4(\frac{1}{\sqrt{1-\frac{\sin^2\alpha}{n^2}}}\frac{\cos\alpha}{n})=2-4(\frac{\cos\alpha}{\sqrt{n^2-\sin^2\alpha}})=0$$ (1.53)

From (1.53) follows

$$\frac{\cos\alpha}{\sqrt{n^2-\sin^2\alpha}}=\frac{1}{2}\qquad 4\cos^2\alpha=n^2-\sin^2\alpha$$

$$4-4\sin^2\alpha=n^2-\sin^2\alpha\qquad \sin\alpha=\sqrt{\frac{4-n^2}{3}}$$

Hence

$$\alpha^* = \arcsin\sqrt{\frac{4-n^2}{3}} = 40.1°$$

The value of the second derivative of δ is

$$\frac{d^2\delta}{d\alpha^2}(\alpha^*) = 1.357$$

Then $\alpha_{min} = \alpha^*$ and from (1.50) follows $\beta = 22.8°$ and from (1.52) $\delta = 168.9°$.

1.2.13 A moving mirror

A plane mirror M (Fig. 1.48) rotates clockwise around an axis normal in

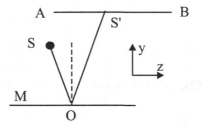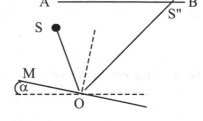

Fig. 1.48

O the plane yz of the figure. A source (having a fixed position in S of coordinates $y = 30$ cm, $z = -5$ cm) emits a beam of light that is reflected on O and impinges on the point S' of a screen AB distant $h = 100$ cm from M. Calculate the coordinates of the real image S' and of the corresponding virtual image S_0 below the mirror.

After the rotation of the mirror M of an angle $\alpha = 2°$, without moving the source S and the screen AB calculate the new coordinates of the real image S" and of the corresponding virtual image S_2 below the mirror. Determine the equation of the plane figure passing through the three points S, S_0 and S_2.

Solution:
From coordinates of S (Fig. 1.49) we have for β, without regard to its sign

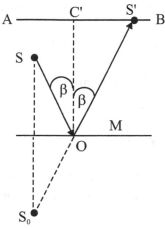

Fig.1.49

$$\beta = \arctan \frac{5}{30} = 9.46°$$

The coordinates of S' are (0, 100, 16.7) cm, because

$$C'S' = h \tan \beta = 100 \cdot \tan(0.1651) = 16.7\text{cm}$$

From geometry (Fig. 1.50)

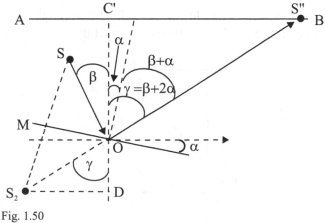

Fig. 1.50

$$\gamma = \beta + 2\alpha = 13.46°$$

hence S" has coordinates (0, 100, 23.9) cm because

$$C'S''=h\tan\gamma=100\cdot\tan(0.2350)=23.9\,\text{cm}$$

The coordinates of virtual image S_0 are (0,-30,-5)
Hence $R = SO = S_0O$

$$R=\sqrt{30^2+5^2}=30.4\text{cm}$$

After rotation there is $R = OS_2 = OS$. The positions S, S_0 and S_2 are points of a circle having radius R, center in O and equation

$$z^2+y^2=R^2=924.16\,\text{cm}^2$$

The coordinates of S_2, are (0, -29.6, -7.1) cm because

$$DS_2=R\sin\gamma=30.4\cdot\sin(0.2350)=7.1\text{cm}$$

$$OD=R\cos\gamma=30.4\cdot\cos(0.2350)=29.6\text{cm}$$

1.2.14 A transparent plate
A beam of monochromatic light is incident with an angle α on a transparent plate (Fig. 1.51) of thickness $h = 1$ cm. If the refractive index is n

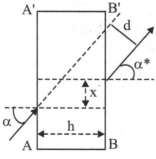

Fig, 1.51

$= 1.45$ determine α^*, x and d when $\alpha = 2°$, $4°$, $6°$, ..., $60°$. If the refractive index is a linear function of the distance between AA' and BB' with values $n' = 1.45$ on AA' and $n'' = 2.5$ on BB' determine the corresponding values of x. Plot as a function of α the values x and d if n is fixed and if n is varying linearly. Give in the two conditions the values of x and d for $\alpha = 40°$.

Solution:
When the refractive index is fixed we have

$$\sin\alpha = n\sin\beta \qquad \beta = \arcsin(\frac{1}{n}\sin\alpha) \qquad (1.54)$$

and from geometry (Fig. 1.52)

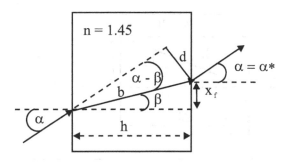

Fig. 1.52

$$b = \frac{h}{\cos\beta} \qquad x = h\cdot\tan\beta \qquad d = b\cdot\sin(\alpha-\beta)$$

The refracted ray from the first face is the reflected ray on the second face, hence

$$n\sin\beta = \sin\alpha* \qquad (1.55)$$

From (1.55) and (1.54) follows $\alpha = \alpha*$.
The continuous linear function between AA' and BB' of the refractive index is

$$n(z) = n' - (n''-n')\frac{z}{h}$$

with z varying in the interval $(0, h)$. Assuming $N = 100$ (for example) and

$$\Delta n = \frac{n'' - n'}{N}$$

we have the following discrete values of n

$$n_1, n_2, ..., n_i, n_{i+1}, ..., n_N$$

with

$$n_{i+1} = n_i + \Delta n_1 \qquad i = 1, 2, ..., N-1 \qquad n_1 = n'$$

Assuming also $\Delta h = h/N$ we have (Fig. 1.53)

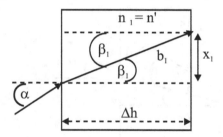

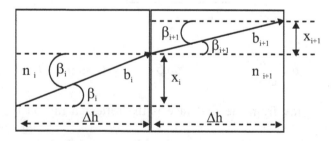

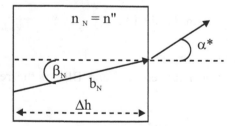

Fig. 1.53

$$\sin\alpha = n_1 \sin\beta_1 = ... = n_i \sin\beta_i = ... = n_N \sin\beta_N = \sin\alpha$$

and for every β_i

$$x_i = \Delta h \times \tan\beta_i \qquad b_i = \frac{\Delta h}{\cos\beta_i} \qquad d_i = b_i \times \sin(\beta_{i-1} - \beta_i)$$

The final values will be (Fig. 1.54)

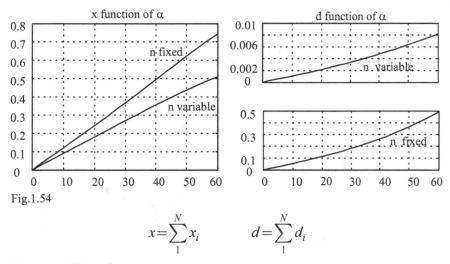

Fig.1.54

$$x = \sum_1^N x_i \qquad d = \sum_1^N d_i$$

For $\alpha = 40°$ we have

Table 1.3

$\alpha = 40°$	n fixed	n variable
x (mm)	4.9	3.6
d (mm)	2.6	0.1

1.2.15 A wave guide

A beam of monochromatic light ($\lambda = 0.65\ \mu$) is subjected in free space ($n = 1$) to multiple reflections between the two plane parallel mirrors S and S' distant h (Fig. 1.55). It is required that the wave incident in A and the

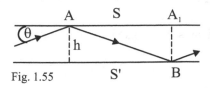

Fig. 1.55

wave reflected in B must have the same phase.

Define the N angles θ_m which satisfy this condition. Plot θ_m as a function of m with $h = 5$ μ.

Find how N varies with h for a fixed value of λ.

With $h = 5$ μ and the minimum angle θ_m, determine the light velocity along the direction A-A$_1$, the distance AB and the time necessary to light to move from A to B.

Solution:

The passage from A to B, and not to C, in the interval Δt, is due to reflection on A. From geometry (Fig. 1.56)

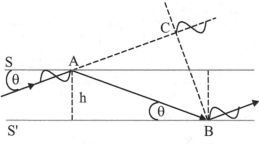

Fig. 1.56

$$AB = \frac{h}{\sin\theta} \qquad\qquad AC = AB\cos 2\theta$$

The path difference is

$$s = AB - AC = AB(1 - \cos 2\theta) = 2h\sin\theta$$

and the corresponding phase difference, assuming two additional phase shifts of π due to the reflection on A and B, is

$$\varphi = 2\pi + k(AB - AC) = 2\pi + \frac{2\pi}{\lambda} 2h\sin\theta$$

The wave in B replicates the phase of wave in A if

$$2\pi + \frac{2\pi}{\lambda} 2h\sin\theta = p2\pi$$

where p is an integer. Hence

$$\frac{2h\sin\theta}{\lambda}=p-1 \tag{1.56}$$

with $p = 2, 3, ...$ For $p = 1$ there isn't reflection because $\theta = 0$. The (1.56) can be written

$$2h\sin\theta=m\lambda \qquad m=1, 2, 3,... \qquad m=p-1 \tag{1.57}$$

Rounding we have $N = 2h/\lambda = 15$, the maximum value of m, for $\sin\theta = 1$. For a fixed λ, N is a linear function of h.

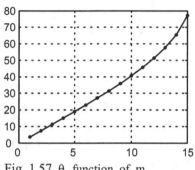

Fig. 1.57 θ_m function of m

From (1.57) the angles θ_m are (Fig. 1.57 and Table 1.4)

$$\theta_m=\arcsin(\frac{m\lambda}{2h}) \qquad m=1,2,...,15$$

Table 1.4

m	1	2	3	4	5	...	11	12	13	14	15
$\theta_m(°)$	3.7	7.5	11.2	15.1	19.0	...	45.6	51.3	57.7	65.5	77.2

With $h = 10\mu$ and $\theta_m = 3.7°$, the light speed from A to A_1 is

$$v=\frac{s}{t}=\frac{AA_1}{AB/c}=c\frac{AA_1}{AB}=c\frac{AB\cos\theta_m}{AB} = c\times\cos\theta_m = 2.99\times10^8\text{m/sec}$$

The paths AB and AA_1 are (in micron)

$$AB=\frac{h}{\sin\theta}=76.9 \qquad AA_1=AB\cos\theta=76.8$$

and the light travels from A to B in the time $t_{AB}=AB/c=0.3$ picosec

1.2.16 Two mirrors
Between two flat parallel mirrors, S and S' (Fig. 1.58), separated by a di-

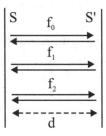

Fig. 1.58

stance $d=15$ cm and with a coefficient ρ (ratio of reflected intensity to the incident intensity) equal to 0.95, there is free space ($n=1$). A thin beam of light, perpendicular to S and S', is reflected back and forth. The light has a wavelength varying from 0.4 to 0.7 μ. Dividing this interval with a step of 3 nm, find the values of λ which satisfy the condition of standing waves calculating and plotting their relative intensities

Solution:
If the wavelengths λ have to be standing waves with d fixed, must be (Fig. 1.59) $d=m(\lambda/2)$ with m integer, or $m=2d/\lambda$.
For example, if $d=15$cm and $\lambda=0.5$μ would be $m=600000$. Scanning, with a step of 3nm, values of λ the corresponding values of m can be calculated. Defining with ε the ratio of the amplitude of the reflected to incident wave,

$$\varepsilon=\sqrt{\rho}$$

(The reflectivity ρ in Sec. 3.1.3 is called R^*)

$$\varphi=2\frac{2\pi}{\lambda}d=\frac{4\pi d}{\lambda} \qquad\qquad (1.58)$$

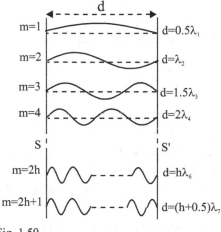

Fig. 1.59

the phase change after a round trip of the ray between S and S', and α the initial phase of f_0 we have for the first wave, considered harmonic

$$f_0 = a\sin\alpha = ae^{i\alpha}$$

and for the following ones

$$f_1 = \varepsilon a e^{i(\alpha+\varphi)} = \varepsilon e^{i\varphi} f_0 \quad f_2 = \varepsilon^2 e^{i2\varphi} f_0 \quad \dots \quad f_n = \varepsilon^n e^{in\varphi} f_0$$

The resulting wave will be

$$f = f_0\left(1 + \varepsilon e^{i\varphi} + \varepsilon^2 e^{i2\varphi} + \dots + \varepsilon^n e^{in\varphi}\right)$$

and the corresponding relative intensity (see Appendix 2)

$$I_r = \frac{1}{1+\varepsilon^2 - 2\varepsilon\cos\varphi} \tag{1.59}$$

The maximum of (1.59) is

$$I_{r\,max} = \frac{I_0}{(1-\varepsilon)^2}$$

for

$$\varphi = 2m\pi \qquad\qquad (1.60)$$

Using (1.58) and (1.60) we have again the initial condition

$$\frac{4\pi d}{\lambda} = 2m\pi \qquad d = m\frac{\lambda}{2}$$

In the range $(0.4, 0.7)$ μ , using a step of 3 nm, we have N values of λ and of the corresponding value of m.

From the N values of m are picked out the N_1 values m_1 that are integer or near to an integer and the corresponding N_1 values of λ_1.

The corresponding N_1 values of φ_1, given by the (1.58), and of I_{r1}, given by (11.59), are calculated.

The N_1 values of λ_1, m_1, φ_1 and I_{r1} are given in the Table 1.5. The values of I_{r1} as a function of λ_1 are plotted in Fig. 1.60.

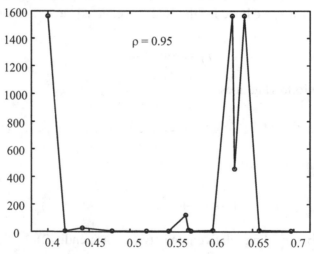

Fig.1.60 Relative intensity for λ in the range $(0.4-0.7)$ μ

Table 1.5

$\lambda\ (\mu)$	m	$\varphi\ (°)$	I_r
0.400	**750000**	**0**	**1560**
0.421	712589	26.5	1.3
0.442	678733	11.4	6.5
0.478	627615	22.6	1.7
0.520	576923	27.7	1.2
0.547	548446	25.0	1.4
0.568	528169	5.07	32.2
0.571	525394	16.4	3.2
0.574	522648	30.1	1.0
0.601	499168	19.2	2.4
0.625	**480000**	**0**	**1560**
0.628	477707	2.29	145.3
0.640	**468750**	**0**	**1560**
0.658	455927	18.6	2.5
0.697	430416	24.8	1.5

1.2.17 Three mirrors

Three plane mirrors are placed in free space ($n = 1$) as in Fig. 1.61.

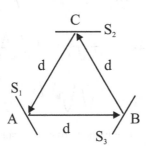

Fig. 1.61

Assume that each mirror has a coefficient $\rho = 0.95$ (see Sec. 1.2.16) and introduces a phase shift of π. The point A is distant $d = 15$cm from B, and so B from C and C from A. Also assume that a beam of light, represented as a harmonic wave, is reflected back and forth on the mirrors without escaping. Find the relation between d and λ if the waves have to be considered standing. If λ is specified in the interval (0.4 - 0.7) μ with a step of 0.5 nm calculate the values of λ that determine a maximum of relative intensity I_r of the reflected light. Plot the highest 20 values of I_r versus the related λ.

Solution:

The distance d must be (Fig. 1.62, Fig. 1.63 and Fig. 1.64) equal to an odd integer number of $\lambda/2$

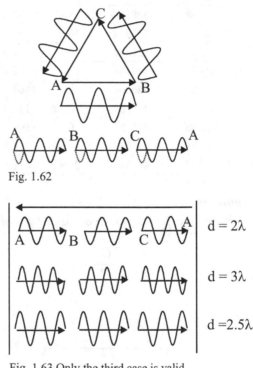

Fig. 1.62

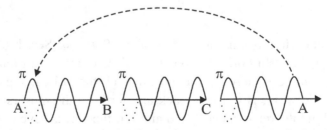

Fig. 1.63 Only the third case is valid

Fig.1.64 The additional shift of π due to the reflection is displayed

$$d = (2m_1 + 1)\frac{\lambda}{2} \qquad m_1 = 0, 1, 2, \dots$$

Through a roundtrip must be

$$g = 3d = 3(2m_1 + 1)\frac{\lambda}{2} = (2m + 1)\frac{\lambda}{2} \tag{1.61}$$

with $m = (3m_1 + 1)$ or, (Table 1.6), $m = 1, 4, 7, \dots$.

Table 1.6

m_1	0	1	2	3
$d\,(\lambda/2)$	1	3	5	7
m	1	4	7	10
$g\,(\lambda/2)$	3	9	15	21

Starting from A the beam is represented by a wave function f_0 with initial phase α

$$f_0 = a \sin \alpha = a e^{i\alpha}$$

After the first running along the path A-B-C the wave function is

$$f_1 = a\varepsilon e^{i(\alpha + \varphi)} = f_0 \varepsilon e^{i\varphi}$$

with ε defined in Sec. 1.2.16, and a change in phase (Fig. 1.65)

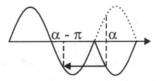

Fig. 1.65

$$\varphi = kg - 3\pi \qquad \longrightarrow \qquad \varphi = kg - \pi \tag{1.62}$$

and for the following passages along A-B-C

$$f_2 = f_0 \varepsilon^2 e^{i2\varphi} \qquad \dots \qquad f_n = f_0 \varepsilon^n e^{in\varphi}$$

Their sum is

$$f = f_0(1+\varepsilon e^{i\varphi} + \varepsilon^2 e^{i2\varphi}...+\varepsilon^n e^{in\varphi})$$

and the corresponding relative intensity (see Appendix 2)

$$I_r = \frac{I}{I_0} = \frac{1}{1+\varepsilon^2 - 2\varepsilon\cos\varphi} \qquad (1.63)$$

The (1.63) has the maximum

$$I_{r\,max} = \frac{1}{(1-\varepsilon^2)}$$

for

$$\varphi = 2m\pi \qquad (1.64)$$

Using (1.62) and (1.64) we get back again the (1.61)

$$kg - \pi = 2m\pi \qquad \frac{2\pi}{\lambda}g - \pi = 2m\pi \qquad g = (2m+1)\frac{\lambda}{2}$$

In Fig. 1.66 the requested relative intensity is plotted as a function of λ.

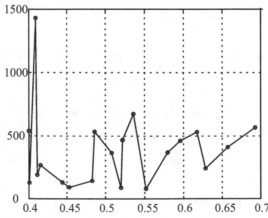

Fig. 1.66 Relative intensity function of λ (μ)

For λ = 0.409 μ we find the highest value of relative intensity (I_r = 1431) and the corresponding values of m (1100244) and φ (360°).

1.2.18 Cavity 1

Between two plane mirrors S and S', separated by a distance $d = 0.5$ m a beam of light, normal to the mirrors, is bouncing back and forth in free space (fig. 1.67). The light has a wavelength in the interval (6328 ± 0.016)

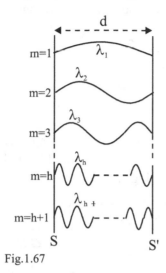

Fig.1.67

Å. S and S' have a coefficient of reflexivity $\rho = 0.95$ (su Sec. 1.2.16). Define the λ_i and the corresponding m_i for which the condition of standing waves $d = m_i(\lambda_i/2)$ holds giving the plot of the relative intensity $I_{r\,i}$. Determine the positions between S and S' where $I_{r\,i}$ has a maximum value and the elapsing time between two maxima of $I_{r\,i}$ in the same position

Solution:

1. Table 1.7 gives the first pair of λ_i and m_i when the condition of standing waves holds for a fixed d.

Table 1.7

$m = 1$	$\lambda_1 = 2d$
$m = 2$	$\lambda_2 = 2d/2$
$m = 3$	$\lambda_3 = 2d/3$
....	....
$m = h$	$\lambda_h\ 2d/h$

The relative intensities are given by (see Sec. 1.2.16)

$$I_{ri} = \frac{1}{1 + \varepsilon^2 - 2\varepsilon \cos \varphi_i}$$

where ε is the square root of ρ (see Sec. 1.2.16) and

$$\varphi_i = 2k_i d = \frac{4\pi}{\lambda_i} d \tag{1.65}$$

The intensity has a maximum when

$$\varphi_i = 2m_i \pi \tag{1.66}$$

Using (1.65) and (1.66) the condition of standing waves is retrieved. Numerically are searched the N pair of λ_i and m_i and the maxima of I_{ri} are calculated. Table 1.8 and Fig. 1.68 give the results.

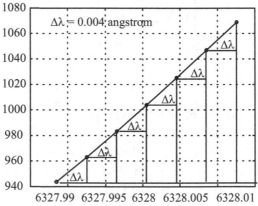

Fig. 1.68 The highest values of I_r function of λ

Table 1.8

λ_i (Å)	m_i	φ_i (°)	I_{ri}
6327.988	1580281	358.8	944
6327.992	1580280	358.8	963
6327.996	1580279	358.9	983
6328.000	1580278	358.9	1004
6328.004	1580277	358.9	1025
6328.008	1580276	359.0	1047
6328.012	1580275	359.0	1069

From Table 1.8 (first column) and Fig. 1.68 we see that the difference between λ_{i+1} and λ_i is always 0.004 Å.

If a standing wave λ_i can be represented as harmonic plane wave it has the form

$$y = a\cos\omega t \sin k_i x \qquad k_i = \frac{2\pi}{\lambda_i}$$

with x a point of an axis normal to S and S'. The standing wave y has its amplitude $A = a\cos\omega t$, varying with t, on the positions

$$k_i x = (2p+1)\frac{\pi}{2} \qquad x_p = (2p+1)\frac{\lambda_i}{4} \qquad p = 0,1,2,...,p_{max}$$

For $p = 0$ and $\lambda_1 = 6327.988$ Å, the first position of A (in micron) is

$$x_0 = \frac{\lambda_1}{4} = \frac{6327.988}{4} = \frac{0.63}{4} = 0.16$$

The amplitude A has a maximum when

$$\omega t = n2\pi \qquad t = nT = n\frac{\lambda}{c} \qquad n = 1,2,3,...$$

For n = 1 and $\lambda_1 = 6327.988$ Å, the first time for the maximum of amplitude A is

$$t = T = \frac{\lambda}{c} = 2.1 \cdot 10^{-15} \sec$$

On the position $x_0 = 0.16$ μ, and the next x_p points, a peak, having a maximum of A and a relative intensity 944 (Table 1.8) and (Fig. 1.69), will occur about every two femtosec.

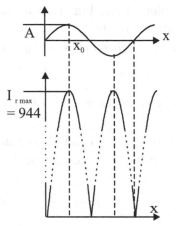

Fig.1.69

1.2.19 Cavity 2

Between two plane mirrors S and S', separated by a distance $d = 0.5$m a beam of light, normal to the mirrors, is bouncing back and forth in free space ($n = 1$). The light has a wavelength in the interval (6328 ± 0.016) Å. S and S' have a coefficient of reflexivity $\rho = 0.95$ (see Sec. 1.2.16). Define the bandwidth Δv of the frequency of light in the assigned interval and the number N of frequencies v_i corresponding to the λ_i for which holds the condition of standing waves. Give the plot of the relative intensity $I_{r\,i}$ as a function of the v_i.

Solution:

1. The maximum and minimum values and their difference are for λ (in Å)

$$\lambda_{min} = 6327.984 \qquad \lambda_{max} = 6328.016 \qquad \Delta\lambda = 0.032$$

and for v (in Hz)

$$v_{min} = 4.740822 \cdot 10^{14} \quad v_{max} = 4.740846 \cdot 10^{14} \quad \Delta v = 2.4 \cdot 10^{9} \quad (1.67)$$

The standing wave condition as a function of the frequency becomes

$$\lambda_i = \frac{2d}{m_i} \qquad \frac{c}{v_i} = \frac{2d}{m_i} \qquad v_i = m_i \frac{c}{2d} = m_i \Delta F$$

with ΔF (in Hz)

$$\Delta F = \frac{c}{2d} = 3 \cdot 10^8 \qquad (1.68)$$

The frequencies, proportional to m_i, are separated by ΔF (Fig. 1.70) that

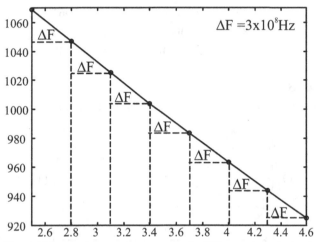

Fig. 1.70 In abscissa are shown only the fifth and sixth decimal digits (see Table 1.9) of the frequency

is the inverse of the time required for a round trip of the light between S and S'.

Using (1.67) and (1.68) we have

$$N = \frac{\Delta v}{\Delta F} = \frac{2.4 \cdot 10^9 \text{ Hz}}{3 \cdot 10^8 \text{ Hz}} = 8$$

The relative intensity is (see Sec. 1.2.18 also for definition of ε and φ_i)

$$I_{ri} = \frac{1}{1 + \varepsilon^2 - 2\varepsilon \cos \varphi_i}$$

The eight best values of the frequencies and related values are calculated using a numerical procedure (Table 1.9 and Fig. 1.70).

Table 1.9

m_i	$v_i (10^{14}\text{Hz})$	$\varphi_i (°)$	$I_{r\,i}$
1580275	4.740825	359.0	1069
1580276	4.740828	359.0	1047
1580277	4.740831	358.9	1025
1580278	4.740834	358.9	1004
1580279	4.740837	358.9	983
1580280	4.740840	358.8	963
1580281	4.740843	358.8	944
1580282	4.740846	358.8	925

1.2.20 Cavity 3

Two concave spherical mirrors A and B (Fig. 1.71) are set in free space

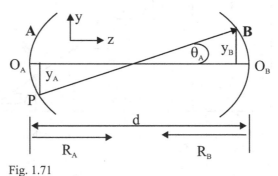

Fig. 1.71

with a distance d from O_A to O_B. Their radii are R_A = -10 mm and R_B = 10 mm. A beam of parallel light, considered a paraxial ray, is reflected back and forth on A and B. Assume that the first ray starts from P (distant y_A from the optical axis O_A - O_B) and makes an angle θ_A = 5° with the optical axis. Determine the range of values of d that inhibits the escaping of the ray outside the mirrors. This is called condition of resonance. Known a value of d, that fulfills the resonance condition, define the N allowed values of the pair y_A and θ_A that permits the condition of resonance.

Solution:

The passage of the ray from A to B (Fig. 1.72**C2**) takes on the matrix form

(see Sec. 1.1.4)

$$T = \begin{vmatrix} y_B \\ \theta_A \end{vmatrix} = \begin{vmatrix} 1 & -d \\ 0 & 1 \end{vmatrix} \begin{vmatrix} y_A \\ \theta_A \end{vmatrix} \tag{1.69}$$

that allows to ascertain the point where the ray hit the mirror B, $y_B = T(1)$. The ray is reflected from the mirror B with an angle θ_B (Fig. 1.72**C3**)

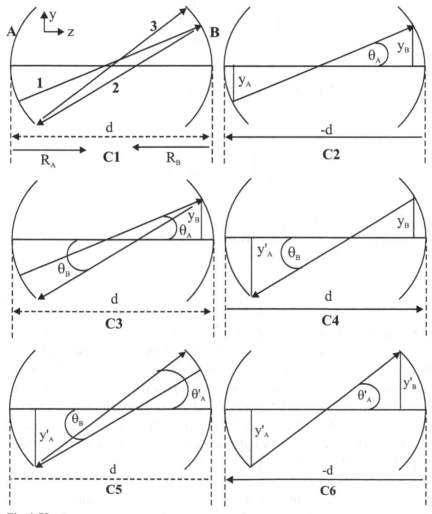

Fig. 1.72

$$S = \begin{vmatrix} y_B \\ \theta_B \end{vmatrix} = \begin{vmatrix} 1 & 0 \\ \dfrac{1}{f_B} & -1 \end{vmatrix} \begin{vmatrix} y_B \\ \theta_A \end{vmatrix}$$

whose value, $\theta_B = S(2)$, is given by the previous matrix equation.
For the ray traveling from B to A we write (Fig. 1.72**C4**)

$$X = \begin{vmatrix} y'_A \\ \theta_B \end{vmatrix} = \begin{vmatrix} 1 & d \\ 0 & 1 \end{vmatrix} \begin{vmatrix} y_B \\ \theta_B \end{vmatrix}$$

and get $y'_A = X(1)$.
Then the ray is reflected on A with an angle θ'_A (Fig. 1.72**C5**). The following equation

$$W = \begin{vmatrix} y'_A \\ \theta'_A \end{vmatrix} = \begin{vmatrix} 1 & 0 \\ \dfrac{1}{f_A} & -1 \end{vmatrix} \begin{vmatrix} y'_A \\ \theta_B \end{vmatrix}$$

We have $\theta'_A = W(2)$.
For the ray traveling from A to B we can write (Fig. 1.72**C6**)

$$Z = \begin{vmatrix} y'_B \\ \theta'_A \end{vmatrix} = \begin{vmatrix} 1 & -d \\ 0 & 1 \end{vmatrix} \begin{vmatrix} y'_A \\ \theta'_A \end{vmatrix} \tag{1.70}$$

and have another value, $y'_B = Z(1)$, of the point where the ray hits the mirror B. The final equation (1.70) has the same form as the first one (1.69). If the initial values θ_A, R_A and R_B are known, a first value of y_A is R_A $\tan\theta_A$. With a value assigned to d is an easy task the recursive multiplication of the previous matrices (Sec. 1.1.4) to obtain subsequent pair (y_B θ_B), (y_A θ_A), etc. Using a MATLAB program the pair can be calculated and plotted (Fig. 1.73) in a straightforward way varying d in a reasonable range between 5 mm ($R_A/2$) and 30mm ($3 R_A$).
From (Fig. 1.73) it follows that if d is between 14 mm and 19 mm the values of θ_A and θ_B are stable below 5°. If we assume $d = 15$ mm, for example, we have repeatedly (Fig. 1.74) the following three values for

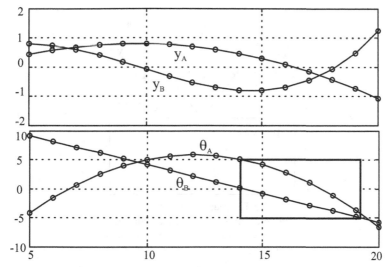

Fig. 1.73 The distance d is in abscissa

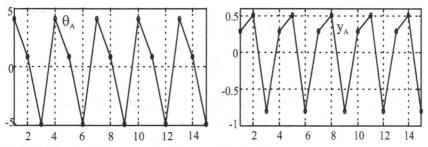

Fig.1.74 The first 15 consecutive θ_A an dy_A for d = 15mm

y_A = 0.2910 mm, 0.5090 mm, -0.8000 mm
and for
θ_A = 4.1673°, 0.8327°, -5.0000°

1.2.21 Fresnel formulae

A beam of monochromatic and linearly polarized light (Fig. 1.75) is incident with an angle θ_1 on the boundary surface between free space and a perfectly transparent medium (n = 1.6). The electric vector oscillates in the plane defined by coordinates wy for the incident ray and by w'y for the reflected ray; the y axis is normal to the plane of figure where lie the axes x, z, w,u, w' and u'. Assume that for the incident ray the amplitude

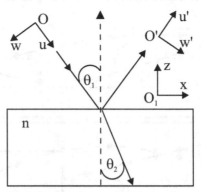

Fig.1.75 The plane of incidence O_1xz
and the y axis are shared equally by
the three coordinate systems xyz, wuy
and w'u'y

of the electric vector is $E = 1.4142$ and the angle between the electric vector and the plane of incidence is $\alpha = 45°$. Hence the components of amplitude of the electric vector E parallel to the w and y axis are

$$E_w = E_y = 1$$

The Fresnel formulae give for the reflected ray the magnitude of the components of the electric vector E' parallel to the w' and y axis

$$E'_{w'} = \frac{\tan(\theta_1 - \theta_2)}{\tan(\theta_1 + \theta_2)} \tag{1.71}$$

$$E'_y = -\frac{\sin(\theta_1 - \theta_2)}{\sin(\theta_1 + \theta_2)} \tag{1.72}$$

Define and plot the magnitude of the electric vector E' and the angle α' between the direction of E' and the w' axis when θ_1 varies in the interval $(0, 90°)$. Discuss the condition and value of phase difference between incident and reflected wave.

Solution:
The angle of refraction θ_2 is a function of θ_1

$$\theta_2 = \arcsin(\frac{\sin\theta_1}{n})$$

Because the incident ray propagates from free space to a medium of refractive index $n>1$, is always $\theta_1 > \theta_2$. Hence is always $E'_y < 0$. On the contrary $E'_w > 0$ if $\theta_1 + \theta_2 < 90°$ and $E'_w < 0°$ (Fig. 1.76) if $\theta_1 + \theta_2 > 90°$.

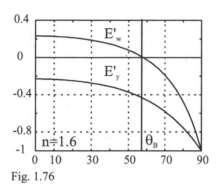

Fig. 1.76

For $\theta_1 + \theta_2 = 90°$, $E'_w = 0$ and the angle of incidence becomes $\theta_B = \arctan(n)$ which is known as Brewster angle.

Varying θ_1 in the interval $(0, 90°)$ and using (1.71) and (1.72) the magnitude E' of the electric vector

$$E' = \sqrt{(E'_{w'})^2 + (E'_y)^2}$$

and the angle α' (between E' and w')

$$\alpha' = \arctan\frac{E'_y}{E'_{w'}}$$

can be calculated and plotted (Fig. 1.77 and Fig. 1.78).

The amplitude of the electric vector E' changes from $E' = 0.3264$ for $\theta_1 = 0°$ to $E' = 1.4142$ when $\theta_1 = 90°$ (Fig. 1.77 and Fig. 1.79).

The angle α' between the direction of the electric vector E' and the w' axis changes from $\alpha' = -45°$ to $\alpha' = -135°$ (Fig. 1.78 and Fig. 1.79).

When θ_1 varies in the interval $(0, 90°)$ the values of $E (= 1.4142)$ and α $(= 45°)$ of the incident ray remain constant but E' and α' of the reflected ray change.

E and *E'* are the amplitude of waves oscillating with incident and reflected ray. If the second medium is optically denser than the first, because, whichever the values of θ_1, *E'* and α', *E* and *E'* have different signs the phase of the reflected wave differ by 180° from the phase of the incident wave. The phase remains equal if the second medium is optically less dense than the first.

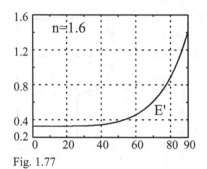

Fig. 1.77

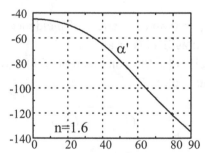

Fig. 1.78 α' is the angle between E' and the w axis on the plane wy varying the angle of incidence

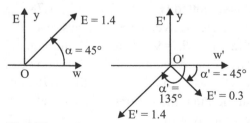

Fig.1.79

Chapter 2

Lenses

2.1 Main Laws and Formulae

2.1.1 Preliminary definitions, restrictions and conventions

Only simple lenses with spherical surfaces are considered (Fig. 2.1). Their radii are R_1 and R_2. Their centers C_1 and C_2 define a straight line

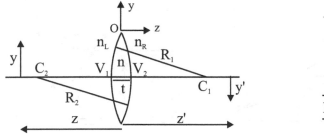

Fig. 2.1   Fig. 2.2 The focus F'

called optical axis of the lens. A plane surface through the optical axis defines a coordinate system yz and is the place where the z positions and the y heights of objects and their images are located. The left and right vertices of a lens, V_1 and V_2, are the points of intersection of the spherical surfaces with the optic axis. The lens thickness is

$$t = |V_1 - V_2|$$

A simple lens is thin when t is negligible in comparison with the product $R_1 R_2$ (see the first power formula in Sec. 2.1.2). When t is not negligible the lens is called thick. A compound lens is a combination of two or more lenses having one or more vertices between the lenses coincident and located on the same optic axis. Two or more lenses having between them distances d and sharing the same optic axis form a coaxial system.

The media to the left, inside and to the right of the lens are supposed to be uniform and their refractive indices, n_L, n and n_R, can be different. Usually n_L and n_R are both equal to one.

If a set of rays, parallel to the optic axis, is incident on a thin lens, all the rays (or their geometrical extensions) converge, after passing through the lens, on a point of the optic axis called focus F' (Fig. 2.2). In Fig. 2.3 the

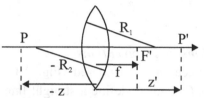
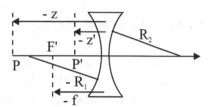

Fig. 2.3

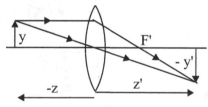
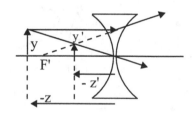

Fig.2.4

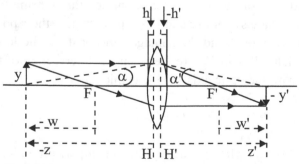

Fig.2.5 The lens is in air

main conventions about the sign are summarized if we assume the light ray is traveling from the left to the right. Fig. 2.4 represents the ray tracing of an object for a biconvex and a biconcave thin lens.

The focal length f is the distance from the vertex to the focus, if the lens is thin (Fig. 2.2); it is the distance from a principal point to the focus if the lens is thick (Fig. 2.5). The distance from vertex to the principal point is denoted with h. For a thick lens, a compound lens and for a coaxial system of two or more lenses their focal lengths are measured from the principal planes. These are not necessarily inside the thick lens or a compound lens or between the lenses of a coaxial system. Besides they can have a direction other than perpendicular to the optic axis.

2.1.2 Formulae for the powers P and focal lengths f

For a thick lens, whose spherical surfaces have radii R_1 and R_2 and made with a material of refractive index n, the power (or lens-maker's) formula is

$$P=\frac{1}{f}=(n-1)(\frac{1}{R_1}-\frac{1}{R_2}+\frac{n-1}{n}\frac{t}{R_1 R_2})$$

that, for a thin lens, becomes

$$P=\frac{1}{f}=(n-1)(\frac{1}{R_1}-\frac{1}{R_2})$$

For two coaxial thin lenses in air, having powers P_1 and P_2 and separated by a distance d, the power P is

$$P=P_1+P_2-P_1P_2 d$$

For a coaxial system of N thin lenses in air, of powers $P_1, P_2, ..., P_N$, with a distance d between the lenses, the power P of the system can be determined using a sequence of instructions (a MATLAB script) repeating a specified number of times the matrix algebra of a lens (see Sec. 2.1.5). For a doublet, a compound of two lenses in contact ($d = 0$) and of powers P_1 and P_2, the power is

$$P=P_1+P_2$$

For a compound of N lenses, of powers $P_1, P_2, \ldots, P_N$, in contact between them, the power P is

$$P=\sum_{i=1}^{N}P_i$$

2.1.3 Thick and thin lenses equations: the Gaussian formulae

The distances h and h' (Fig. 2.5) are given by the formulae

$$h=\frac{(1-n)tf}{nR_2} \qquad h'=\frac{(1-n)tf}{nR_1}$$

where all variables have already been defined for a thick lens. When $t = 0$ there is also $h = h' = 0$ and the distances of object and image from a thin lens are measured from vertices rather than from principal points.

The formula that connects the focal lengths, f and f', to the object and image distances, z and z', when the medium surrounding the lens to the left is different from that to the right ($n_L \neq n_R$) is

$$\frac{f}{z}+\frac{f'}{z'}=1$$

with the power P given by

$$\frac{n_R}{f'}=-\frac{n_L}{f}=P$$

The transverse and angular magnifications are

$$M_T=\frac{n_L}{n_R}\frac{z'}{z} \qquad M_\alpha=\frac{\alpha'}{\alpha}=\frac{z}{z'}$$

When the lens is in air we have only one focal length f (positive or negative as calculated in Sec. 2.1.2) and the previous formulae become

$$f=\frac{1}{P}$$

$$\frac{1}{z'}=\frac{1}{z}+\frac{1}{f}$$

$$M_T = \frac{z'}{z} \qquad M_\alpha = \frac{\alpha'}{\alpha} = \frac{1}{M_T}$$

2.1.4 Thick and thin lenses equations: the Newtonian formulae

The formula that connects the focal lengths (f and f') to the object and image distances, w and w' (see Fig. 2.5 for definitions of the distances w and w'), when the medium surrounding the lens to the left is different from that to the right ($n_L \neq n_R$) is

$$ww' = ff'$$

and the power P

$$-\frac{n_L}{f} = \frac{n_R}{f'} = P$$

The transverse magnification is

$$M_T = -\frac{f}{w} = -\frac{w'}{f'}$$

When the lens is in air we have only one the focal length f (positive or negative as calculated in Sec. 2.1.2) and the previous formulae become

$$f = \frac{1}{P}$$

$$ww' = -f^2$$

$$M_T = \frac{f}{w} = -\frac{w'}{f}$$

2.1.5 Spherical and chromatic aberration

The transverse and the longitudinal spherical aberrations are shown in Fig. 2.6. For correction of chromatic aberration in some simple cases we remember that the refractive index of a transparent material is also

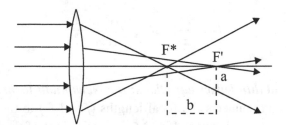

Fig. 2.6 F* is the focus of the marginal rays, F' the
focus of paraxial rays, b the longitudinal and a the
transverse spherical aberration

dependent from the wavelength of the light. So a lens has different focal
lengths for different wavelengths of the light and the position and the
height of the image of an object wary as functions of the wavelength. Two
coaxial lenses, made of the *same* type of glass and having the *positive*
focal lengths f_1 and f_2, must be separated by a distance d

$$d = \frac{1}{2}(f_1 + f_2)$$

to obtain an achromatic combination.

If the two lenses, made of the *different* type of glass, must form a doublet
with assigned focal length f, the achromatic correction requires the following peculiar values for the focal lengths of the lenses

$$f_1 = f\frac{D_2 - D_1}{D_2} \qquad f_2 = -f\frac{D_2 - D_1}{D_1}$$

where D_1 and D_2 are called dispersive powers of the lenses. The definition of the dispersive power is

$$D = \frac{n_F - n_C}{(n_D - 1)}$$

where n_F, n_C and n_D are the refractive indices for some standard wavelengths as, for example, 4861 Å, 5893 Å and 6593 Å usually called the
Fraunhofer F, D and C lines. The value of D is always positive, hence
from the formulae for f_1 and f_2 we have

$$\frac{f_1}{f_2} = -\frac{D_1}{D_2}$$

with the minus sign meaning that the two lenses of an achromatic doublet must be neither both *positive* nor both *negative*.

2.1.6 The matrix form for a thin lens

The Gaussian formula of a thin lens (Fig. 2.7) can relate angles rather than distances

$$\frac{1}{z'} = \frac{1}{z} + \frac{1}{f} \quad \rightarrow \quad \frac{\theta'}{y} = \frac{\theta}{y} + \frac{1}{f} \quad \rightarrow \quad \theta' = \theta + \frac{y}{f}$$

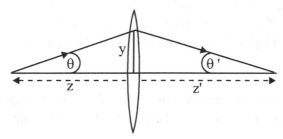

Fig. 2.7

So the latter previous formula can also be obtained from the product of the matrix and vector on the right side of the following equality

$$\begin{vmatrix} y \\ \theta' \end{vmatrix} = \begin{vmatrix} 1 & 0 \\ \frac{1}{f} & 1 \end{vmatrix} \begin{vmatrix} y \\ \theta \end{vmatrix}$$

Using this simple matrix form most of the previous physical quantities are easily calculated for a thin lens and a coaxial system of thin lenses. The matrix form of a refracting surface (see Sec. 1.1.4) is used for determination of properties of a thick lens.

2.2 Problems

2.2.1 Simple thin biconvex lens

A thin double convex lens has a focal length f (Fig. 2.8). With the object to the left of the lens (where distances are negative and objects real) and then with the object to the right of the lens (where distances are positive and objects virtual) determine the values of z' and m_T for the following

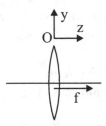

Fig. 2.8

values of z -∞, $-2f$, $-f$, $-f/2$, 0, $f/2$, f, $2f$, ∞. Using the ray tracing draw the position, the state (real or virtual) and the height of the image for the conditions A $(z < -f)$, B $(-f > z < 0)$, C $(0 > z < f)$ and D $(z > f)$.
Then with $f = 15$ plot the functions z' and m_T for $-2f \le z \le 2f$.

Solution:
In Table 2.1 we have the values of z' and of m_T for the requested values of z. Fig. 2.9 and Fig. 2.10 give the graphical requested representations.

Table 2.1

z	-∞	$-2f$	$-f$	$-f/2$	0	$f/2$	f	$2f$	∞
z'	f	$2f$	∞	$-f$	0	$f/3$	$f/2$	$2f/3$	f
m_T	0	-1	-∞	2	0	4/6	3/6	2/6	0

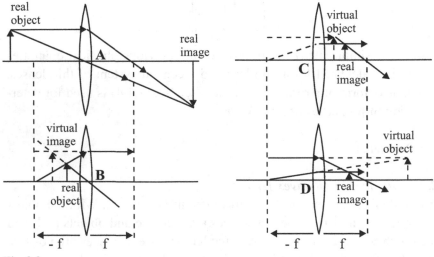

Fig. 2.9

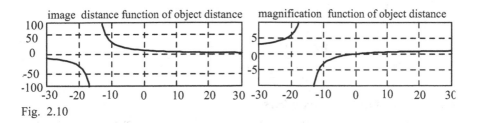

image distance function of object distance   magnification function of object distance

Fig. 2.10

2.2.2 Simple thin biconcave lens

The lens has focal length $f = -140$ mm. First consider an object at a distance $z = -160$ mm and after at a distance $z = 160$ mm (Fig. 2.11). In both

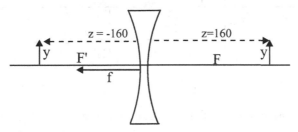

Fig. 2.11

cases draw the ray tracing, locate and calculate the z, z' (Gaussian notation), the w, w' (Newtonian notation) distances and the associated transverse magnifications.

Solution:

The ray tracing is given in Fig. 2.12 and the requested numerical values are in Table 2.2.

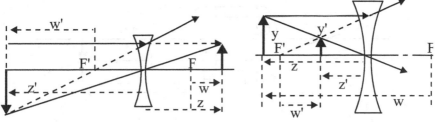

Fig. 2.12 Figures for z =160 mm (left) and z = -160 mm (right)

Table 2.2 (z, z', w, w' are in mm)

z	z'	w	w'	m_T
160	-1120	20	-980	-7
-160	-74.7	-300	65.3	0.5

2.2.3 Simple thick biconvex lens

A thick lens has the following figures: $n = 1.65$, $t = 3$ mm, $R_1 = 35$ mm and $R_2 = -60$ mm (Fig. 2.13). Find the values of f, h and h'. Consider two

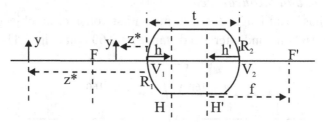

Fig. 2.13

positions for the distance of the object from the vertex V_1: first $z^* = 50$ mm, then $z^* = 30$ mm. For each hypothesis draw the ray tracing, locate and calculate the z, z' (Gaussian notation), the w, w' (Newtonian notation) distances and the associated transverse magnifications.

Solution:

Using formulae of Sec. 2.1.2 and 2.1.3 we have $f = 34.4$ mm, $h = 0.7$ mm from the vertex V_1 and $h' = 1.2$ mm from the vertex V_2. The ray tracing is in Fig. 2.14 for $z^* = -50$ mm and in Fig. 2.15 for $z^* = -30$ mm. The requested numerical values are in Table 2.3.

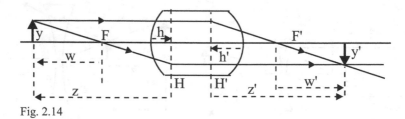

Fig. 2.14

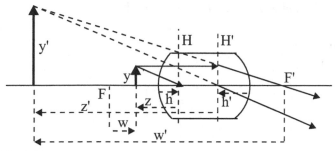

Fig. 2.15

Table 2.3 (z^*, z, z', w, w' are in mm)

z^*	z	z'	w	w'	m_T
-50	-50.7	107.4	-16.2	73.0	-2.7
-30	-30.7	-281.1	3.8	-315.5	9.2

2.2.4 Thin lens 1

A thin positive lens L (Fig. 2.16) has a focal length f = 20 cm with the

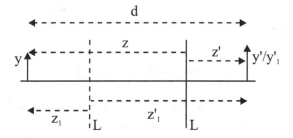

Fig. 2.16

object at a distance z from L and y = 5 cm high. The image, y' high, is at a distance z' from L and d = 100 cm from object. An image in the same position and distant d from the object is obtained shifting L along the optic axis. The distances from L of object and image become z_1 and z'_1 and the image is now y'_1 high. Find z, z_1, z', z'_1, the corresponding pair y' and y'_1 and the relation between d and f if the image has to be real. Demonstrate that

$$y = \sqrt{y' y'_1}$$

and locate the heights of the images using the ray tracing.

Solution:

The basic law

$$\frac{1}{z'}=\frac{1}{z}+\frac{1}{f}$$

since

$$d=z'-z \qquad z'=d+z$$

becomes

$$\frac{1}{d+z}=\frac{1}{z}+\frac{1}{f} \qquad \rightarrow \qquad z^2+dz+df=0$$

The solutions of this quadratic equation

$$\frac{-d\pm\sqrt{d^2-4df}}{2}=\begin{cases} z=-\,72.4\,\text{cm} \\ z_1=-27.6\,\text{cm} \end{cases}$$

are real because

$$d^2-4df>0 \quad \rightarrow \quad d>4f$$

Therefore the distance d must be at least four times the focal length f. The corresponding distances of the images are

$$z'=d+z=27.6\,\text{cm} \qquad z_1'=72.4\,\text{cm}$$

and

$$d=z'_1-z_1 \qquad d=-(z+z_1) \qquad d=z'+z'_1$$

So the sum of the two distances of the object and those of the image are both equal to d and $z' = -z_1$ and $z'_1 = -z$. The transverse magnification in the two conditions is

$$M = \frac{z'}{z} \qquad M_1 = \frac{z'_1}{z_1}$$

or

$$y' = M y \qquad y'_1 = M_1 y$$

Then

$$MM_1 = \frac{z'}{z} \frac{z'_1}{z_1} = \frac{-z_1}{z} \frac{-z}{z_1} = 1$$

and

$$y' y'_1 = y^2 \, MM_1 = y^2 \quad \rightarrow \quad y = \sqrt{y' y'_1}$$

For the ray tracing of the two images see Fig. 2.17.

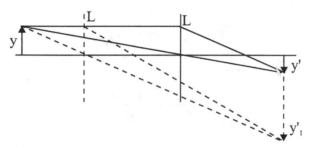

Fig. 2.17

2.2.5 Thin lens 2
The object and the image are located (Fig. 2.18) to the left and to the right

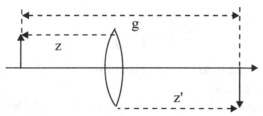

Fig. 2.18

of a thin biconvex lens whose power is $P = 0.05$ cm^{-1}. With g equal to the sum of the object and of the image distance and M_T the transverse magnification, find the function $g = g(M_T)$ and the value M_T* for which there is the minimum g* of the function $g = g(M_T)$. For $g = g$* find the object" and image distances z* and z'*.

Solution:

The formula of a thin lens with

$$g = z' - z \qquad z' = g + z \qquad (2.1)$$

becomes

$$\frac{1}{g+z} = \frac{1}{z} + P \qquad (2.2)$$

From the formula of the transverse magnification we have

$$M_T = \frac{g+z}{z} \qquad z = \frac{g}{M_T - 1} \qquad (2.3)$$

Using (2.3), (2.2) becomes

$$\frac{1}{g + \dfrac{g}{M_T - 1}} = \frac{M_T - 1}{g} + P$$

or

$$g = \frac{1}{P}(2 - M_T - \frac{1}{M_T})$$

The first derivative is

$$\frac{dg}{dM_T} = \frac{1}{P}(\frac{1}{M_T{}^2} - 1)$$

and its critical value is M*$_T$ = -1. The value M*$_T$ = 1 is discarded in the condition of this problem.

Because the second derivative is positive for M*$_T$ = -1, the function g has a minimum for the critical value (Fig. 2.19 and Fig. 20)

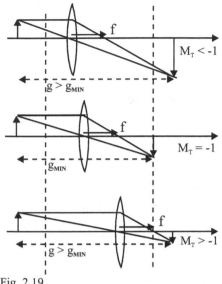

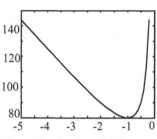

Fig. 2.19   Fig. 2.20 g funtion of M

$$g* = \frac{4}{P} = 80\,\text{cm}$$

From (2.1) and (2.2) using the value of $g*$ it follows

$$z* = \frac{g*}{M_T - 1} = \frac{80}{-2} = -40\,\text{cm} \qquad z'* = g* + z* = 40\,\text{cm}$$

2.2.6 Thick lens

A lens, whose diameter is $h = 23$ mm and focal length $f = 11.1$ mm, must be produced using a glass whose refractive index is $n = 1.993$. The two spherical surfaces will have the same radius R and a thickness t. Find R and t (Fig. 2.21).

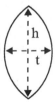

Fig. 2.21

Solution:

The lens-maker's formula is

$$\frac{1}{f}=(n-1)(\frac{1}{R_1}-\frac{1}{R_2}+\frac{n-1}{n}\frac{t}{R_1 R_2})$$

With $A = n-1$, $B = A/n$, $R_1 = R$ and $R_2 = -R$ we have

$$\frac{1}{f}=A(\frac{2}{R}+\frac{Bt}{R^2})$$

that, with $x = 1/R$, we have

$$\frac{1}{f}=A(2x+Btx^2) \quad \rightarrow \quad Bf\,tx^2+2Af\,x-1=0$$

And with

$$a = Btf \qquad b = 2Af \tag{2.4}$$

the initial formula becomes

$$ax^2+bx-1=0 \tag{2.5}$$

From geometry (Fig. 2.22) it follows

$$(\frac{h}{2})^2+(R-\frac{t}{2})^2=R^2 \qquad t^2-4Rt+h^2=0$$

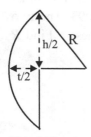

Fig.. 2.22

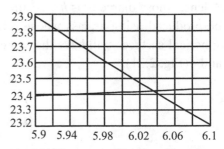

Fig. 2.23 The two set of R values (formulae (2.5) and (2.6)) function of t

and

$$R = \frac{t^2 + h^2}{4t} \qquad (2.6)$$

Assuming t defined in the interval $(1, h/2)$ mm with a step of 0.1 mm, two set of R are calculated using both (2.5) and (2.6). The thickness for which the same value of R occurs, using a MATLAB script, is $t = 6.0$ mm. Hence it follows (Fig. 2.23) $R = 23.4$ mm from both (2.5) and (2.6).

2.2.7 Thin lens 3

A cylinder (Fig. 2.24), filled with water ($n_1 = 4/3$), is closed with a thin biconvex lens ($f = 30$ mm) to the left and with a mirror S to the right. The distance between lens and mirror is $d = 80$ mm. A real object is located at a distance $z = 90$ mm from the lens. Find the distance of the image from the lens and its transverse magnification.

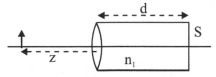

Fig. 2.24

Solution:

When the refractive indices are different on the two sides of the lens the following formulae must be applied

$$\frac{n_1}{f_1} = -\frac{n}{f} = P \qquad \frac{f}{z} + \frac{f_1}{z_1} = 1$$

where f (negative) is the focal length to the left of the lens and f_1 (positive) the focal length to its right. It follows

$$f_1 = -n_1 f = 40 \, \text{mm}$$

and

$$z_1 = \frac{z f_1}{z - f} = 30 \, \text{mm}$$

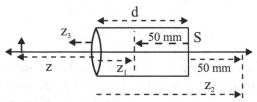

Fig. 2.25

The real image (Fig. 2.25) of the lens becomes real object for the mirror. The virtual image of the mirror becomes real object for the lens located at a negative distance $z_2 = -(50 + 80) = -130$ mm .The distance is now negative because the light travels from the right to the left. So now it is $f_1 = -40$ mm and $f = 30$ mm. The image for the lens will be at a distance z_3 according to the formula

$$\frac{f}{z_3} + \frac{f_1}{z_2} = 1$$

hence

$$z_3 = \frac{z_2 f}{z_2 - f_1} = 22.9 \text{ mm}$$

The final transverse magnification will be

$$M_T = \frac{z_3 z_1}{z_2 z} = 0.006$$

The final image and the initial object are both located to the left of the lens. But the height of the final image is very smaller than that of the object.

2.2.8 Lens and mirror

An object, $y = 5$ mm high, is located at a distance $z = -61$ mm to the left of a positive thin lens L of focal length $f = 33$ mm. To its right is placed a plane mirror S inclined of 45° with respect to the optic axis of the lens. The point of intersection of the mirror with the optic axis is the focus F' of the lens (Fig. 2.26). At a vertical distance $AF' = d = 22$ mm from the optic axis is situated the plane surface of a box filled with water ($n' = 1.33$). Find the position and the size of the final image.

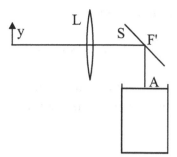

Fig. 2.26

Solution:
The image (Fig. 2.27), y' tall, in the absence of the mirror S, would be located at the distance

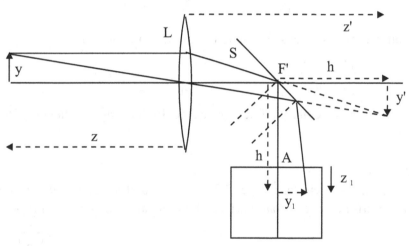

Fig. 2.27

$$z' = \frac{zf}{z+f} = 71.9\,\text{mm}$$

and have the height

$$y' = M_T\, y = \frac{z'}{z}\, y = -5.9\,\text{mm}$$

If the mirror was in place but in absence of the water box the image,

$y_1 = y'$, would be located at a distance, normal to the optic axis

$$h = z' - f = 38.9\,\text{mm}$$

But this image, y_1 long, becomes virtual object for the plane refractive surface A of the water. The distance of this virtual object from the surface A is

$$z_1 = h - d = 16.9\,\text{mm}$$

The distance of the final image is given (see Secs. 1.1.3 and 1.2.3) by the formula ($n = 1$, $n' = 1.33$)

$$\frac{n'}{z_1'} = \frac{n}{z_1} \qquad z_1' = n'z_1 = 22.5\,\text{mm}$$

Because $y_1 = y'$ the transverse magnification can be written

$$\frac{y'_1}{y_1} = \frac{nz'_1}{n'z_1} \quad \longrightarrow \quad y'_1 = \frac{z'_1}{n'z_1}\,y_1 = \frac{z'_1}{z'_1}\,y_1 = y_1 = y'$$

The mirror and the plane refracting surface don't change the height of the image.

2.2.9 Doublet

An object, $y = 2$ cm long, (Fig. 2.28) is placed at a distance $h = 10$ cm from a mirror S, whose plane surface makes an angle of 45° with the optic

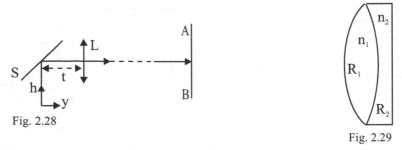

Fig. 2.28

Fig. 2.29

axis of a doublet lens L. The point of intersection of the mirror with the optic axis is distant $t = 5$ cm from L. The image must be projected on the screen AB, distant 5 m from L. The first thin lens (Fig. 2.29) of the doublet has $n_1 = 1.5097$ and $D_1 = 0.0155$ and the second lens has $n_2 = 1.8503$ and $D_2 = 0.0311$. D_1 and D_2 are the dispersive powers of the two materials forming the lenses (see Sec. 2.1.5). Draw the ray tracing of the system assuming L as a single thin lens. Find the focal length of doublet L and the radii R_2 and R_1 if the doublet has to be achromatic.

Solution:
In Fig. 2.30 we have the ray tracing for the mirror (left) and for L (right) considered as a thin lens.

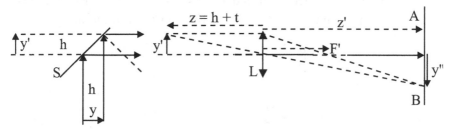

Fig. 2.30

The distance of the virtual object for the doublet is $z = h + t = -15$ cm
If $z' = 50$ m, then

$$f = \frac{zz'}{z-z'} = \frac{-15 \times 500}{-15 - 500} = 15.6 \, \text{cm}$$

If the doublet has to be achromatic, for the first lens the focal length must be

$$f_1 = f \frac{D_2 - D_1}{D_2} = 7.30 \, \text{cm}$$

and for the second lens

$$f_2 = f \frac{D_2 - D_1}{D_1} = -14.65 \, \text{cm}$$

Since is $R'_2 = \infty$ from the lens-maker's formula

$$\frac{1}{f_2} = (n_2 - 1)(\frac{1}{R_2} - \frac{1}{R'_2})$$

it follows $R_2 = f_2(n_2-1) = -12.46$ cm. And with $R_2 = R'_1$ from

$$\frac{1}{r_1} = \frac{1}{f_1(n_1-1)} + \frac{1}{R'_1} = a = 0.19 \text{ cm}^{-1}$$

it follows $R_1 = 5.3$ cm.

2.2.10 Lens and prism

The optical system is composed of a transparent prism, whose section is a right triangle with AB = BC = h = 8 cm, and a thin positive lens distant s = 20 cm from the face BC of the prism (Fig. 2.31). The lens has a focal length f = 15 cm. An object is placed at a distance t = 20 cm from the surface AB of the prism (n = 1.5). Find the final position of the image.

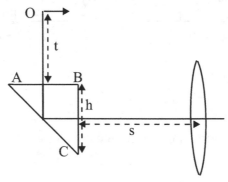

Fig. 2.31

Solution:

A ray from O traveling normal to the surface AB of the prism will undergo a total reflection on D (Fig. 2.32). Indeed the critical angle for total reflection is for this prism 42°. Everywhere a ray impinges normally to the surface AB the path of light in the prism is always h = 8 cm. Hence

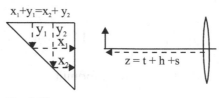

Fig. 2.32

the distance of the object from the lens is $z = t + h + s = -38$ cm and the corresponding distance of the image to the right of the lens is

$$z' = \frac{zf}{z+f} = 24.8 \, \text{cm}$$

2.2.11 Coaxial system 1

A coaxial system is made up by the positive lens L_1 ($f_1 = 60$ cm) and the negative one L_2 ($f_2 = -15$ cm). Focus to the right of L_1 and that to the right of L_2 occupy the same position (Fig. 2.33). A beam of parallel ray reaches the lens L_1 with a deviation α from the optic axis. Give the ray tracing,

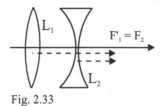

Fig. 2.33

the position of the final image, the angular magnification and the power of the system. Verify the power of the system when the distance t between L_1 and L_2 varies in the interval (10 - 80) cm.

Solution:

The ray tracing is given in Fig. 2.34. If the lens L_2 wasn't be present the image would be located in A (Fig. 2.34). With L_2 in place the image made

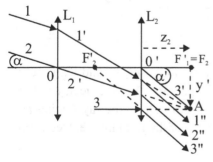

Fig. 2.34 Rays 3 and 3' define the new direction of rays 1" and 2"

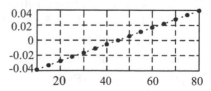

Fig.2.35 Power P (cm^{-1}) function of t (cm)

by L_1 becomes a virtual object for L_2. The final image is a parallel beam of light whose direction makes an angle α' with the optic axis. So the distance z'_2 of the final image is at infinite.
From Fig. 2.34 we have

$$\tan\alpha=\frac{AF_2}{OF_2} \qquad \tan\alpha'=\frac{AF_2}{O'F_2}$$

and the angular magnification is

$$M_\alpha=\frac{\tan\alpha'}{\tan\alpha}=\frac{OF_2}{O'F_2}=\frac{f_1}{f_2}=4$$

The power P of the system is

$$P=P_1+P_2-t\,P_1\,P_2=0.0167+(-0.067)-(-0.050)=0$$

where t is the distance between the two lenses and P_1 and P_2 are the powers of the first and second lens respectively.
Obviously $f=\infty$ and then $P=0$ are immediate consequence of the previous result $z'_2=\infty$.
When the distance t varies in the interval (10 - 80) cm, the values of the power P (cm^{-1}) are given in Fig. 2.35 where we find $P=0$ for $t=45$ cm. The corresponding values of the focal length are given in Table 2.4.

Table 2.4

t (cm)	10	20	30	40	45	50	60	70	80
f (cm)	-25.7	-36	-60	-180	∞	180	60	36	25.7

2.2.12 Coaxial system 2
The distance between two positive and coaxial lenses, L_1 and L_2, is equal to the focal distance ($f_2=15$ cm) of L_2 (Fig. 2.36). A real object, $y=4$ cm high, is placed in the focus of L_1 ($f_1=30$ cm). Draw the ray tracing and find the height y' and the distance z' from L_2 of the final image. Determine the power P of the system.

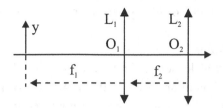

Fig. 2.36

Solution:

Draw first (Fig. 2.37) the line 5 that changing direction when crosses L_2 becomes the line 6. The line 4, parallel to the line 5, doesn't change direction crossing L_2. The line 1 leaving L_1 must have the same direction of the line 5 and leaving L_2 must converge at the point of intersection of lines 6 and 3.

We have, remembering the sign convention (Sec. 1.1.2)

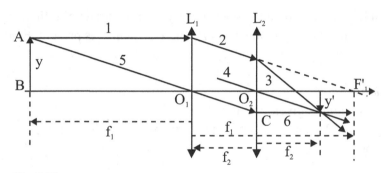

Fig. 2.37

$$\frac{y}{y'} = \frac{f_1}{-f_2} \quad \rightarrow \quad y' = \frac{-f_2}{f_1} y = -2 \text{ cm}$$

and $z' = f_2 = 15$ cm.

From the formula of the power P

$$P = P_1 + P_2 - t\, P_1 P_2$$

substituting the focal distances to the powers and f_2 to t, we have

$$\frac{1}{f} = \frac{1}{f_1} + \frac{1}{f_2} - \frac{t}{f_1 f_2} = \frac{1}{f_1} + \frac{1}{f_2} - \frac{f_2}{f_1 f_2} = \frac{1}{f_2}$$

Hence the focal distance of the system is the same as that of L_2.

2.2.13 Coaxial system 3

Two thin and coaxial lenses L_1 and L_2 are distant $d = 15$ cm and have focal lengths $f_1 = 20$ cm and $f_2 = -10$ cm. A beam of parallel rays crossing the optic axis with an angle $\alpha = 3°$ enters L_1 (Fig. 2.38). If L_2 wasn't present a point image would appear, to the right of L_1, y'_1 high from optic

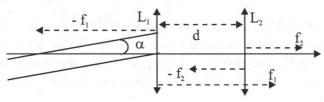

Fig. 2.38

axis and z'_1 distant from L_1. Define the ray tracing of the point final image and the corresponding final values z'_2 and y'_2 when L_2 is present. Find the power P of the system.

Solution:

The ray tracing is given in Fig. 2.39.
As $z'_1 = f_1 = 20$ cm and $y'_1 = 20 \tan 3° = 1.0$ cm, we have $z_2 = f_1 - d = 5$ cm. From

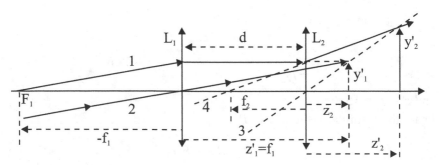

Fig. 2.39

$$\frac{1}{z'_2} = \frac{1}{z_2} + \frac{1}{f_2}$$

it follows $z'_2 = 10$ cm.

From law of transverse magnification

$$y'_2 = \frac{z'_2}{z_2} y'_1$$

it follows $y'_2 = 2$ cm.

The power P is

$$P = P_1 + P_2 - t\, P_1 P_2 = 0.05 + (-0.1) - (0.075) = 0.025\,\text{cm}^{-1}$$

and $f = 40$ cm.

2.2.14 Coaxial system 4

A beam of light parallel to the optic axis enters a system of two thin and coaxial lenses L_1 and L_2 and goes out from L_2 again with the same direction (Fig. 2.40a). The distance between L_1 and L_2 is $d = 15$ cm. Shifting only L_2 to the right of $b = 20$ cm the image becomes a point located on the optic axis and distant $a = 56.3$ cm from the new position of L_2 (Fig. 2.40b). Find the focal lengths f_1 an f_2.

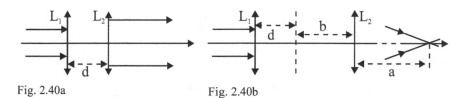

Fig. 2.40a   Fig. 2.40b

Solution:

Supposing first L_1 negative and L_2 positive we have $z_1 = \infty$ and $z'_1 = f_1$ (f_1 negative). If rays go out of L_2 (Fig. 2.41a) parallel to the optic axis, it must be

$$f_2 = d - f_1 \qquad z_2 = -f_2 \qquad z'_2 = \infty$$

with f_2 positive.

Shifting L_2 to the right (Fig. 2.41b) the virtual point object for L_2 is

$$z_3 = -(d+b-f_1)$$

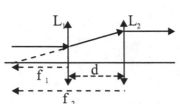

Fig. 2.41a   Fig. 2.41b

From law

$$\frac{1}{z'_3} = \frac{1}{z_3} + \frac{1}{f_2}$$

using the previous value of z_3 and assuming $z'_3 = a$ we have

$$\frac{1}{a} = \frac{1}{-(d+b-f_1)} + \frac{1}{d-f_1} \quad \rightarrow \quad \frac{1}{a} = -\frac{1}{b+x} + \frac{1}{x}$$

with

$$x = d - f_1 \qquad z'_3 = a$$

that is

$$x^2 + bx - ab = 0$$

with $b = 20$ cm and $ab = 1125$ cm^2
The solutions are

$$x = d - f_1 = f_2 = \begin{cases} +25\,\text{cm} \\ -45\,\text{cm} \end{cases}$$

hence

$$f_1 = \begin{cases} d-x=15-25=-10\,\text{cm} \\ d-x=15+45=+60\,\text{cm} \end{cases}$$

Therefore can be either L$_1$ negative with L$_2$ positive ($f_1 = -10$ cm and $f_2 =$

25 cm, see Fig. 2.42a) or L_1 positive with L_2 negative ($f_1 = 60$ cm and $f_2 = -45$ cm, see Fig. 2.42b).

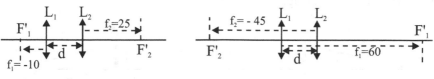

Fig. 2.42a   Fig. 2.42b

2.2.15 Coaxial system 5

Two thin and coaxial lenses L_1 and L_2 have focal distances $f_1 = 20.0$ mm and $f_2 = 10.0$ mm. A beam of parallel rays is incident on L_1 forming an angle α with optic axis (Fig. 2.43). The rays go out from L_2 parallel forming an angle α^* with optic axis. Find the angular magnification and the power of the system.

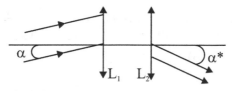

Fig. 2.43

If the higher ray of the incident beam (Fig. 2.44) crosses the focus to the left of L_1 with an angle $\alpha = 10°$, find the least values of $2a_1$ and $2a_2$ that allow the refraction from both the lenses of all incident rays and the corresponding values d_1 and d_2 of the beams incident and emerging from the lenses.

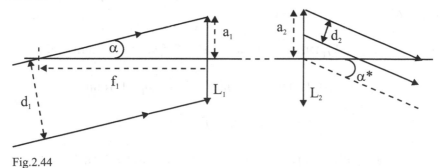

Fig.2.44

Solution:

The distance between the lenses must be equal to the sum of the focal lengths of L_1 and L_2 (Fig. 2.45). Hence

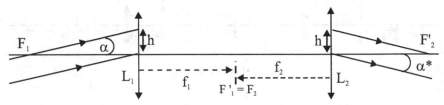

Fig. 2.45

$$\alpha = \frac{h}{-f_1} \qquad \alpha^* = \frac{h}{f_2} \qquad M_\alpha = \frac{\alpha^*}{\alpha} = -\frac{f_1}{f_2} = -2.0$$

For the power the obvious result is zero because $f = \infty$. We can check this from the formula

$$P = (\frac{1}{f_1} + \frac{1}{f_2} - \frac{t}{f_1 f_2}) = (\frac{1}{f_1} + \frac{1}{f_2} - \frac{f_1 + f_2}{f_1 f_2}) = 0$$

Also we have (Fig. 2.46)

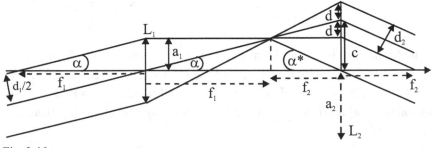

Fig. 2.46

$$2a_1 = 2f_1 \tan \alpha = 7.1 \text{mm} \qquad d_1 = 2f_1 \sin \alpha = 6.9 \text{mm}$$

$$c = (f_1 + f_2) \tan \alpha = 5.3 \text{mm} \quad d = c - a_1 = 1.8 \text{mm}$$

$$\alpha^* = \arctan(a_1 / f_2) = 19.5°$$

$$2a_2 = 2(c + d) = 14.2 \text{mm} \qquad d_2 = 2d \cos \alpha^* = 3.4 \text{mm}$$

2.2.16 Principal planes

A beam of rays parallel to the optic axis is incident on the first lens L_1 (f_1 = 81.3 cm) of a coaxial system. The other lens L_2 (f_2 = 87.8 cm) is distant $t = 37$ cm from the first one. Find the focal length f of the system and the distances of the principal planes H and H' from L_1 and L_2. Determine these values also for $t_1 = 71.4$ cm and $t_2 = 178.6$ cm.

Solution:

The power P and the focal length f of the system are

$$P = P_1 + P_2 - t P_1 P_2 = 0.0185 \, \text{cm}^{-1} \qquad f = 1/P = 54.0 \, \text{cm}$$

Considering only L_1 (Fig. 2.47) we would have a real point image at the position

$$z'_1 = f_1 = 81.3 \, \text{cm}$$

that becomes a virtual point object for L_2 at distance z_2. The real point image F for L_2 is at distance z'_2 from this. Their distances are

$$z_2 = f_1 - t = 44.3 \, \text{cm} \quad z'_2 = \frac{f_2 z_2}{z_2 + f_2} = 29.4 \, \text{cm} \quad h' = f - z'_2 = 24.6 \, \text{cm}$$

The point F is really the position of the focus of the system of the two lenses: its distance from L_2 is 29.4 cm but focal distance is $f = 54.0$ cm. The point H' distant 54.0 cm from F is a *principal point* and the *principal plane* intersects normally in H' the optic axis.

Exchanging the positions of L_1 and L_2 in the system (Fig. 2.48) we have

$$z'_1 = f_2 \quad z_2 = f_2 - t = 50.8 \, \text{cm} \quad z'_2 = 31.3 \, \text{cm} \quad h = f - z'_2 = 22.8 \, \text{cm}$$

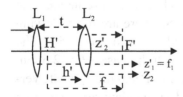

Fig. 2.47

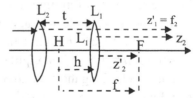

Fig. 2.48

Fig. 2.49 shows the positions of principal planes. The focal distance f is related to value of t so that f is positive if t is less than the sum of f_1 and f_2, becomes infinite when t is equal to this sum and is negative if t is grater than this sum (Fig. 2.50).

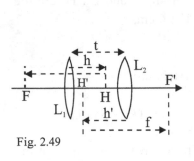

Fig. 2.49

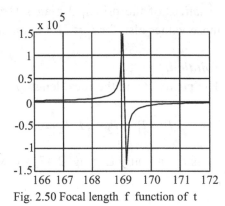

Fig. 2.50 Focal length f function of t

Fig. 2.51 shows values and positions of f, H and H' for the assigned values t_1 and t_2.

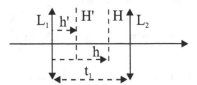

Fig. 2.51a We have (in cm) t_1 =71.4, f = 73.1, h' = 7.2 and h = 59.5

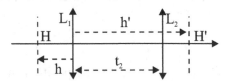

Fig. 2.51b We have (in cm) t_2 = 178.6, f = -753.6, h' = 1833.9 and h = -1532.8

2.2.17 Two lenses and a prism

A monochromatic light source emits a beam that passing through the slit C reaches L_1. The emerging rays from L_1 are parallel and, after undergoing the condition of minimum deviation through a prism, are incident on L_2 that produces a clear image on the screen S. There are the following distances: $a = 8$ cm, $b = 4.5$ cm, $d = 6.5$ cm and $e = 7$ cm (Fig. 2.52). The prism has $n = 1.5$ and an equilateral section with the base $AB = 2.6$ cm. Beam travels along the path DE equidistant from AB and the opposite vertex. Find the angle of deviation δ subtended by incident and emerging

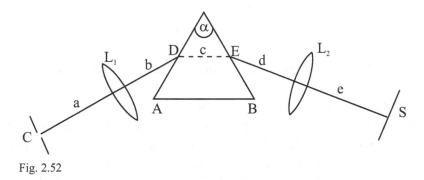

Fig. 2.52

rays through the prism, the focal lengths f_1 and f_2 of the two lenses, the power and the focal length of the system. Draw the ray tracing of the equivalent coaxial system.

Solution:

With n and α defined a value of δ that allow the matching of the two terms of the following equation

$$n\sin(\frac{\alpha}{2})=\sin\frac{\alpha+\delta}{2}$$

is $\delta = 37.2°$. In Fig. 2.53 the horizontal line is the first term and the slanted line the second term of the previous equation. The focal length of L_1 is $f_1 = a = 12$ cm for the rays emerge parallel from lens and is $f_2 = e = 7$ cm since a beam of parallel rays reaches the screen S.

Taking $t = b + c + d = 12.3$ cm ($c = 1.3$ cm from the geometry of the equilateral triangle), the focal length of the system is

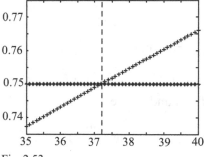

Fig. 2.53

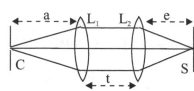

Fig. 2.54

$$\frac{1}{f} = \frac{1}{f_1} + \frac{1}{f_2} - \frac{t}{f_1 f_2} = 0.048\,\text{cm}^{-1} \qquad f = 20.8\,\text{cm}$$

Fig. 2.54 displays the ray tracing of the equivalent coaxial system.

2.2.18 Longitudinal and transverse spherical aberrations

A light beam parallel to the optic axis reaches a circular (radius r_1) sur-face of a planoconvex lens ($R = 20$ mm, $t = 3$ mm, $n = 1.67$). A maximum value for r_1 must be defined sufficient to satisfy the condition of paraxial rays within an error of 2%. Then define the interval (z_{min}, z_{max}) of the lon-gitudinal aberration and, for the transverse aberration, the area of the disk (radius r_2) of the real image (Fig. 2.55). Why a geometrical point can't be a real image?

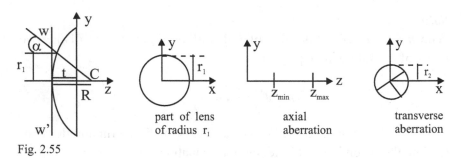

Fig. 2.55

| part of lens of radius r_1 | axial aberration | transverse aberration |

Solution:

Since for α (in radiant) less than 20° there is $\alpha = \sin\alpha = \tan\alpha$ with an error less than 2% it can be assumed $\alpha_{max} = 20°$. Let use y_{max} for r_1 given by (Fig. 2.56)

$$y_{max} = r_1 = R\sin\alpha_{max} = 6.8\,\text{mm}$$

Dividing the interval (0, y_{max}) into $N=21$ values of α (= 0°, 1°, . . ., 20°) (Fig. 2.57), the corresponding values of y, d and α' are

$$y = R\sin\alpha \qquad d = R(1-\cos\alpha) \qquad \alpha' = \arcsin(\frac{\sin\alpha}{n})$$

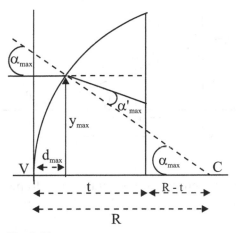

Fig. 2.56

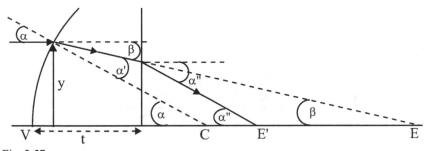

Fig. 2.57

On the plane surface of the lens the rays are incident with an angle β and emerge with α''. Hence

$$\beta=\alpha-\alpha' \qquad n\sin\beta=\sin\alpha'' \qquad \alpha''=\arcsin(n\sin\beta)$$

and (Fig. 2.58)

$$y-y'=(t-d)\tan\beta \qquad y'=y-(t-d)\tan\beta$$

Rays emerging from the lens intersect the optic axis at different point distant z from the plane surface of the lens

$$y'=z\tan\alpha'' \qquad z=y'/\tan\alpha''$$

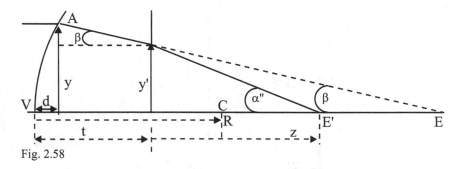

Fig. 2.58

To two of N values y_i and y_{i+1} correspond the quadruplet E'_i , α''_i , z_i, y'_i, and the quadruplet E'_{i+1} , α''_{i+1} , z_{i+1}, y'_{i+1} (Fig. 2.59).

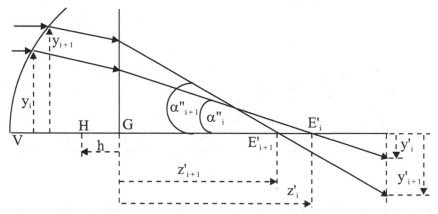

Fig. 2.59

The first and the last ($N = 21$) terms of z'_i and y'_i are $z'_1 = 28.1$ mm, $z'_N = 26.9$ mm and $y'_1 = 0$, $y'_N = -0.28$ mm.
The focus of the lens is

$$P=\frac{1}{f}=\frac{n-1}{R} \qquad f=29.9\,\text{mm}$$

The value z'_1 is surely satisfying the condition of paraxial rays. In fact it corresponds to $\alpha = 0°$. Then for the principal point H we can write

$$h=-(f-z_1)=-1.8\,\text{mm}$$

and for r_2 and for the area D of the image disk of radius r_2

$$r_2 = y''_N = 0.28\,\text{mm} \qquad D = 0.25\,\text{mm}^2$$

The final image must have an area different from zero because the intensity of light cannot be infinite.

2.2.19 Image intensity

A beam of light passes through a circular aperture C (r = 10 mm) of a plane A and reaches the full surface of the lens (f = 80.0 mm and d = 40 mm). The beam crossing the aperture C has an intensity $I = 3.1831 \times 10^5$ watt/m². The circular aperture is z =200 mm distant from the lens. The image of C is formed to the right of the lens on the screen S. Find the power P of the light crossing the circular aperture, the intensity I_L of the beam on the lens and the intensity I_S of the image assuming there isn't reflection on the lens or absorption by this (Fig. 2.60).

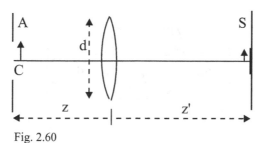

Fig. 2.60

Solution:

The power P at the circular aperture is

$$P = I\pi r^2 = 100\,\text{watt}$$

The intensity of light on the lens is

$$I_L = \frac{P}{\pi \left(\dfrac{d}{2}\right)^2} = 8.0 \times 10^4\,\frac{\text{watt}}{\text{m}^2}$$

The distance between the lens and the screen is

$$z' = \frac{zf}{z+f} = 133.3\,\text{mm}$$

The radius of the image is

$$|r'| = \left|\frac{z'}{z}r\right| = 6.7\,\text{mm}$$

and its intensity

$$I_S = \frac{P}{\pi r'^2} = 7.2 \cdot 10^5 \frac{\text{watt}}{\text{m}^2}$$

2.2.20 Focal length of a thick lens

A thick lens has $R_1 = 20$ mm, $R_2 = -35$ mm, $n_2 = 1.993$ and $t = 5.6$ mm (Fig. 2.61). Find the focal length of the lens and, using matrix algebra, the distances d_1, d_2 and the corresponding positions of the principal planes.

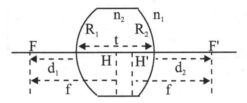

Fig. 2.61

Solution:

The focal distance is

$$\frac{1}{f} = (n_2 - 1)\left(\frac{1}{R_1} - \frac{1}{R_2} + \frac{n_2 - 1}{n_2}\frac{t}{R_1 R_2}\right) \qquad f = 13.5 \text{ mm}$$

Taking (Fig. 2.62) $y_1 = 1$ and $\theta_1 = 0°$ and using matrix calculus for the ray refracted on the first spherical surface (see Sec. 1.1.4)

$$A = \begin{vmatrix} y_1 \\ \theta_2 \end{vmatrix} = \begin{vmatrix} 1 & 0 \\ (1 - \frac{n_1}{n_2})\frac{1}{R_1} & \frac{n_1}{n_2} \end{vmatrix} \begin{vmatrix} y_1 \\ \theta_1 \end{vmatrix} \tag{2.7}$$

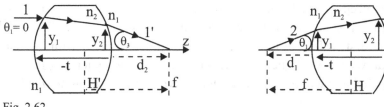

Fig. 2.62

we define

$$\theta_2 = A(2) = 1.43°$$

Then for the ray going from the first to the second spherical surface

$$B = \begin{vmatrix} y_2 \\ \theta_2 \end{vmatrix} = \begin{vmatrix} 1 & -t \\ 0 & 1 \end{vmatrix} \begin{vmatrix} y_1 \\ \theta_2 \end{vmatrix} \tag{2.8}$$

we obtain

$$y_2 = B(1) = 0.86 \, \text{mm}$$

and for the ray refracted on the second spherical surface

$$C = \begin{vmatrix} y_2 \\ \theta_3 \end{vmatrix} = \begin{vmatrix} 1 & 0 \\ (1 - \dfrac{n_2}{n_1}) \dfrac{1}{R_1} & \dfrac{n_2}{n_1} \end{vmatrix} \begin{vmatrix} y_2 \\ \theta_2 \end{vmatrix} \tag{2.9}$$

we have

$$\theta_3 = G(2) = 4.2°$$

and finally

$$d_2 = \frac{y_2}{\tan \theta_3} = 11.6 \, \text{mm} \tag{2.10}$$

So the position of the principal plane H' from the second spherical surface is

$$f - d_2 = 1.9 \, \text{mm}$$

Now we need to find the position of the principal plane H. The ray incident on the first spherical surface, crossing the optic axis ad a distance d_1 from the first surface of the lens, must subtend a value θ_1 with the optic axis that, with the assigned $y_1 = 1$, will make the ray emerging from the lens parallel to the optic axis. If we have a value of d_1 the angle θ_1 will be

$$\theta_1 = \arctan(\frac{y_1}{d_1}) \tag{2.11}$$

The requested value of d_1 is, for our lens, less than f. So trying with a set of N values of d_{1i} within the interval (11.2, 13.5) mm, a corresponding set of N values for θ_{1i} is obtained using (2.11). Iterating, using a MATLAB script, for N times the sequence of operations (2.7) – (2.9) it is found for $d_1 = 12.4$ mm (Fig.2.63)

$$y_2 = 1.1 \text{ mm} \qquad \theta_1 = -4.6° \qquad \theta_3 = 0.0004° = 0°$$

Then the position of the principal plane H to the right of the first spherical surface is

$$f - d_1 = 1.1 \text{ mm}$$

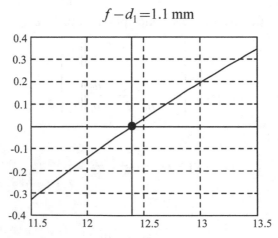

Fig. 2.63 Angle θ_3 function of d_1

2.2.21 Three coaxial lenses

An object is located at a distance $d = 130$ mm to the left of the first of three coaxial lenses of same focal length f and having distance d between

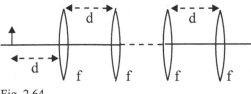

Fig. 2.64

them (Fig. 2.64). To simplify the problem we suppose $f < d$ and f defined in the interval (60, 70) mm; for every value of f find for each lens the positions z_1 of the object, z_2 of the image and the transverse magnification, the final position and final transverse magnification of the image using either the matrix algebra or the recursive use of standard formulae (see Sec. 2.1.3). There is a value of f for which the distance of the final image from the last lens is equal to that of the object from the first one?

Solution:

Using matrix notation (see the left side of Fig. 2.65) for a ray incident on the first lens subtending an angle $\theta_1 = -4.4°$ with the optic axis, we have $y_1 = 10$ mm $(= d \tan\theta_1)$ and

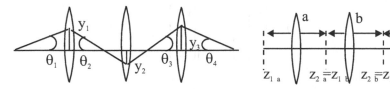

Fig. 2.65

$$A = \begin{vmatrix} y_1 \\ \theta_2 \end{vmatrix} = \begin{vmatrix} 1 & 0 \\ \dfrac{1}{f} & 1 \end{vmatrix} \begin{vmatrix} y_1 \\ \theta_1 \end{vmatrix}$$

with

$$\theta_2 = A(2)$$

Known θ_2 and y_1 we find the distance of the image from the first lens. From the ray traveling from the first to the second lens

$$B = \begin{vmatrix} y_2 \\ \theta_2 \end{vmatrix} = \begin{vmatrix} 1 & d \\ 0 & 1 \end{vmatrix} \begin{vmatrix} y_1 \\ -\theta_2 \end{vmatrix}$$

we can calculate

$$y_2 = B(1)$$

From ray crossing the second lens

$$C = \begin{vmatrix} y_2 \\ \theta_3 \end{vmatrix} = \begin{vmatrix} 1 & 0 \\ \dfrac{1}{f} & 1 \end{vmatrix} \begin{vmatrix} y_2 \\ \theta_2 \end{vmatrix}$$

it follows

$$\theta_3 = C(2)$$

Known θ_3 and y_2 we find the distance of the image from the second lens. From the ray traveling from the second to the third lens

$$D = \begin{vmatrix} y_3 \\ \theta_3 \end{vmatrix} = \begin{vmatrix} 1 & d \\ 0 & 1 \end{vmatrix} \begin{vmatrix} y_2 \\ -\theta_3 \end{vmatrix}$$

we obtain

$$y_3 = D(1)$$

From the ray crossing the third lens

$$E = \begin{vmatrix} y_3 \\ \theta_4 \end{vmatrix} = \begin{vmatrix} 1 & 0 \\ \dfrac{1}{f} & 1 \end{vmatrix} \begin{vmatrix} y_3 \\ \theta_3 \end{vmatrix}$$

we have

$$\theta_4 = E(2)$$

Known θ_4 and y_3 we find the distance of the final image from the third lens.

Using in a recursive way the standard formulae (see Sec. 2.1.3) for z_2 and m_T known z_1 and f (see the right side of Fig. 2.65)

$$z_2 = \frac{f z_1}{f + z_1} \qquad m_T = \frac{z_2}{z_1} \qquad z_1 = z_2 - d$$

the same results are found. For the focal lengths 60 mm and 70 mm the results are given in Table 2.5 and Fig. 2.66. For $f = 65$ mm the distance of the final image from the last lens is equal to that of the object from the first one and the final magnification is, as expected, one (Fig. 2.67). For the focal lengths 60 and 70 mm the final magnification are 0.8 and 1.5.

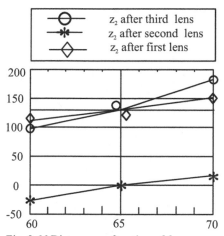

Fig. 2.66 Distances z_2 function of f

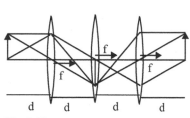

Fig. 2.67

Table 2.5 (Values for f, z_1 and z_2 are in mm)

$f = 60$	1st lens	2nd lens	3rd lens	$f = 70$	1st lens	2nd lens	3rd lens
z_1	-130	-18.6	-156.9	z_1	-130	21.7	-113.4
z_2	111.4	-26.9	97.2	z_2	151.7	16.5	182.8
m_T	-0.9	1.4	-0.6	m_T	-1.2	0.8	-1.6

Chapter 3

Polarization

3.1. Main Laws and Formulas

3.1.1 Linearly polarized light

We assume a right-handed set of axes with the plane *xy* coincident with the page; the beam of light is moving along the direction of the *z* axis and the electric field **E**, normal to the *z* axis, oscillates in the plane *xy only* in a fixed direction (Fig. 3.1) that subtends angle θ with the *x* axis. Dealing with polarization the magnetic field is usually ignored.

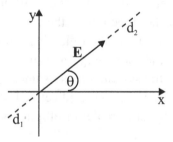

Fig 3.1 The straight line d_1 - d_2 is the direction of vibration of **E** and the transmission axis for a linear polarizer

Representing with a plane harmonic wave the beam of a linearly polarized light the oscillation of **E** is

$$E = E_M \sin(kz + \omega t) = E_M e^{i(kz+\omega t)} = E_M e^{i\varepsilon}$$

107

where E_M is the vectorial value of the peak amplitude. The components of the electric vector in the plane xy are

$$E_x = i\,A_x \sin(kz + \omega t) = i\,A_x e^{i(kz+\omega t)} = i\,A_x e^{i\varepsilon}$$

$$E_y = j A_y \sin(kz + \omega t) = j A_y e^{i(kz+\omega t)} = j A_y e^{i\varepsilon}$$

where now i and j are the unit vectors and

$$A_x = E_M \cos\theta \qquad A_y = E_M \sin\theta$$

the scalar value of the peak amplitude of the components of the electric field along x and y axis.

There are many physical processes (Stark, Raman, Zeeman or Cerenkov effects, molecular scattering of light, *etc.*), where light or non visible radiation appears polarized, which are of theoretical interest only. A *linear polarizer* is the practical device used to produce a linearly polarized light and is made exploiting proprieties of birefringence of some crystals (a Nicol prism, first realized in 1828) and of dichroic microcrystals (a Polaroid sheet, first invented in 1928). We shall call θ also the angle between the x axis and the direction of the *transmission axis*, or *pass-plane*, (Fig. 3.1). We are using, in the following problems, an "ideal" linear polarizer that doesn't absorb energy and gives rise to a "full linear" polarization through the transmission axis. A linear polarizer can also be briefly called polarizer (P) omitting the adjective linear. An adjective will always be used for non linear (circular or elliptic) polarizer.

3.1.2 Elliptically polarized light

Non linear polarized light is obtained sending a linearly polarized beam through a *phase plate*, also called a *retarder*. This has, lying in its plane, a special direction, called *optic axis* (see Sec. 3.1.4 and remember that this phrase has a different meaning in Chapter 2). Assuming a beam of linearly polarized light, moving along the z axis and reaching normally the retarder, situated in the xy plane (Fig. 3.2), the optic axis can make any angle β with the x axis (see Sec.3.1.5). We suppose the retarder is a sheet of a uniaxial crystal, placed in the xy plane, with its *optic axis coinciding with the x axis* and that a beam of linearly polarized light, with the plane of oscillation of the electric field E subtending an angle θ with x axis, reaches normally the retarder (Fig. 3.3). The beam passing through the plate is split into two rays still moving in the z direction but with different velocities (see Sec. 3.1.4 and example b of the next Fig. 3.7). The electric field of first ray, called ordinary, is the y component E_y of E and the electric field of second ray, called extraordinary, is the x component E_x of E.

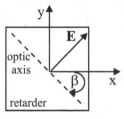

Fig. 3.2

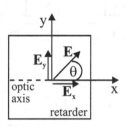

Fig. 3.3

The beam emerging from the plate has a different form of polarization because the x component and y component of the electric field E (Fig. 3.3) are subjected to different phases φ_x and φ_y.
Using scalar quantities for the component of the electric field subtending the angle θ with the x axis, we have

$$E_x = A_x \sin \varepsilon \qquad A_x = E_M \cos \theta$$

$$E_y = A_y \sin(\varepsilon + \varphi) \qquad A_y = E_M \sin \theta$$

where

$$\varphi = \varphi_y - \varphi_x$$

is the difference between the additional phases acquired by the two components.
The previous two formulae are the parametric representation of an ellipse

whose form depends from φ and whose inclination of the major axis depends from θ (Fig. 3.4a and Fig. 3.4b).

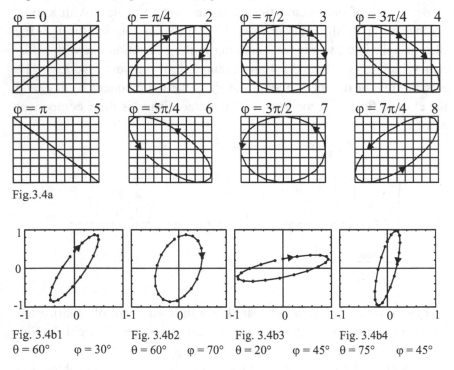

Fig.3.4a

Fig. 3.4b1   Fig. 3.4b2   Fig. 3.4b3   Fig. 3.4b4
$\theta = 60°$ $\varphi = 30°$ $\theta = 60°$ $\varphi = 70°$ $\theta = 20°$ $\varphi = 45°$ $\theta = 75°$ $\varphi = 45°$

Increasing φ with a fixed θ the major axis decreases and the minor increases (Fig. 3.4b1 and Fig. 3.4b2).

Varying θ with a fixed φ (Fig. 3.4b3 and Fig. 3.4b4) varies the angle θ between the major axis and the x axis. When θ approaches the values $0°$ and $180°$ or $90°$ and $270°$ the ellipse becomes a straight line parallel to the x axis or the y axis.

The previous general parametric formulae functions of θ and φ will be used (see Sec. 3.1.5) with matrix calculus. Consider now two simple cases. If the phase difference is $\varphi = \pi$ we have

$$E_x = A_x \sin \varepsilon$$

$$E_y = A_y \sin(\varepsilon + \varphi) = A_y \sin(\varepsilon + \pi) = -A_y \sin \varepsilon$$

On this condition the retarder, called a *half-wave plate* (see graph 5 in Fig. 3.4a), is used to change the inclination of the electric field **E** of a linear form of polarization (from ρ to - ρ if this is the angle between **E** and the optic axis) or to reverse the handedness of a non linear form of polarization.

When $\varphi = \pm\pi/2$ the ellipse assumes the standard form

$$\frac{E_x^{\ 2}}{A_x^{\ 2}} + \frac{E_y^{\ 2}}{A_y^{\ 2}} = 1 \qquad \frac{E_x^{\ 2}}{\cos^2\theta} + \frac{E_y^{\ 2}}{\sin^2\theta} = A_M^{\ 2}$$

and the retarder, called *quarter-wave plate*, produces a form of non linear polarization when θ is different from 0°, 180°, 90° and 270°.

If $\varphi = \pm\pi/2$ and axis of transmission of a linearly polarized beam is placed at $\theta = \pm45°$ from the x axis we have

$$E_x^{\ 2} + E_y^{\ 2} = \frac{E_M^{\ 2}}{2} \qquad A_x = A_y = \frac{E_M}{\sqrt{2}}$$

and the light emerging from the retarder has a circular form of polarization.

With our previous definition of E_x, E_y and φ, if the phase shift is in the interval $(\pi < \varphi < 2\pi)$ the resultant electric field **E** rotates in a *counterclockwise* direction and we speak of a *left polarized* light. If the phase difference is in the interval $(0 < \varphi < \pi)$ the resultant electric field **E** rotates in a *clockwise* direction and we speak of a *right polarized* light.

All the variables in the equations defining the ellipse and the circle are values proportional to the *intensity*.

A device that, as a black box, produces an elliptic or circular polarized light is called an *elliptic or circular polarizer*.

The distinction between a polarizer and a retarder is clear: the former "produces" a form of polarization, the latter "converts" one form of polarization into another one.

3.1.3 Reflectivity and refractivity

We consider a beam of a linearly polarized light moving in the air and incident with an angle α_1 on the plane surface of a transparent substance whose refractive index is n. If the substance is homogeneous and isotropic, the light is represented as a plane and harmonic wave and the electric and magnetic fields are continuous at the boundary surface, the Fresnel equations (see Sec. 1.1.5) gives the component of the electric field in the plane of incidence xz (E_{1w}, $E'_{1w'}$, E_{2w}) and along the y axis (E_{1y}, E'_{1y}, E_{2y}) of the incident, reflected and refracted rays (Fig. 3.5) assuming that the direction of electric field E_1 of the incident ray makes an angle γ different from 90° with the plane wk_1 (Fig. 3.6).

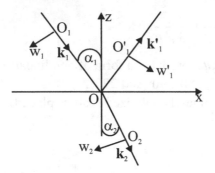

Fig. 3.5 Four right-handed sets of axes (xyz, w_1yk_1, $w'_1yk'_1$, w_2y,k_2) are shown with y-axis direction from the reader to the plane xz

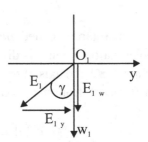

Fig. 3.6 Light moves along k_1 whose direction is out of page from the plane w_1y

Relative intensities of the *component* of the electric field along the axes (w_1, w'_1, w_2) normal to the y axis and along the y axis are derived from Fresnel equations.

$$I_{rw} = \frac{I'_{1w'}}{I_{1w}} = \left|\frac{E'_{1w'}}{E_{1w}}\right|^2 = \frac{\tan^2(\alpha_1 - \alpha_2)}{\tan^2(\alpha_1 + \alpha_2)} \qquad I_{rw} = \left(\frac{n-1}{n+1}\right)^2$$

$$I_{ry} = \frac{I'_{1y}}{I_{1y}} = \left|\frac{E'_{1y}}{E_{1y}}\right|^2 = \frac{\sin^2(\alpha_1 - \alpha_2)}{\sin^2(\alpha_1 + \alpha_2)} \qquad I_{ry} = \left(\frac{n-1}{n+1}\right)^2$$

$$I_{tw} = \frac{I_{2w}}{I_{1w}} = \left|\frac{E_{2w}}{E_{1w}}\right|^2 = \frac{\sin 2\alpha_1 \sin 2\alpha_2}{\sin^2(\alpha_1 + \alpha_2)\cos^2(\alpha_1 - \alpha_2)} \qquad I_{tw} = \left(\frac{2}{n+1}\right)^2 n$$

$$I_{ty} = \frac{I_{2y}}{I_{1y}} = \left|\frac{E_{2y}}{E_{1y}}\right|^2 = \frac{\sin 2\alpha_1 \sin 2\alpha_2}{\sin^2(\alpha_1 + \alpha_2)} \qquad I_{ty} = \left(\frac{2}{n+1}\right)^2 n$$

where α_1 and α_2 are the angles between the incident and refracted rays with the normal to the boundary surface and the subscripts r and t refer to intensities of the reflected and refracted rays. The formulae on the right side above are used when $\alpha_1 = 0°$.

The relative intensities expressed as the ratio of the intensity I_r of the reflected ray by the intensity I_1 of the incident ray and as the ratio of I_t by I_1, called *reflectivity R* and *refractivity T*, are given by the following formulae

$$R = \frac{I_r}{I_1} = I_{rw} \cos^2 \gamma + I_{ry} \sin^2 \gamma$$

$$T = \frac{I_t}{I_1} = I_{tw} \cos^2 \gamma + I_{ty} \sin^2 \gamma$$

with I_{rw}, I_{ry}, I_{tw}, I_{ty} defined in the previous formulae and γ is the angle that the electric field E_1 subtends with the w_1 axis or the plane of incidence (Fig. 3.5 and Fig. 3.6). When $\alpha_1 = 0°$ *reflectivity* and *refractivity* defined by the formulae

$$R = \left(\frac{n-1}{n+1}\right)^2 \qquad T = \frac{4n}{(n+1)^2}$$

are independent from the angle γ.

If reflection and refraction occur without absorption from the substance where they originate, it is always

$$R + T = 1 \qquad \frac{I_r}{I_1} + \frac{I_t}{I_1} = 1 \qquad I_r + I_t = I_1$$

in agreement with principle of conservation of energy.

When $\alpha_1 + \alpha_2 = \pi/2$ we have $I_{rw} = 0$ and the electric field of the reflected light has no component in the plane of incidence wk_1 or xz and the angle of incidence

$$\alpha_{1B} = \arctan n$$

is called Brewster angle.

The degree of polarization for the reflected and refracted rays of polarized light is

$$P_r = \left| \frac{I_{rw} - I_{ry}}{I_{rw} + I_{ry}} \right| \qquad P_t = \left| \frac{I_{tw} - I_{ty}}{I_{tw} + I_{ty}} \right|$$

The *reflectivity* R* and *refractivity* T* for *natural* light are obtained putting $\gamma = 45°$ in the previous formulae for R and T

$$R* = \frac{1}{2}(I_{rw} + I_{ry})$$

$$T* = \frac{1}{2}(I_{tw} + I_{ty})$$

A *natural* (or *completely unpolarized*) light is the visible radiation emitted by a substance that is made to glow by a physics process.

Consider a linear polarized light, moving along the z axis and whose electric field is oscillating in a fixed direction in the plane *xy*. If the beam reaches normally a polarizer, whose pass-plane makes an angle δ with the direction of oscillation of E, the polarizer will transmit a beam whose electric field E_p will be $E\cos\delta$. Hence the corresponding intensities are related by the formula (known as Malus'law)

$$E_p^2 = (E\cos\delta)^2 \qquad I_p = I\cos^2\delta$$

3.1.4 Birefringence

In some crystals, called anisotropic, the refractive index (and then the speed of propagation of the light) is a function of the direction of propagation of the light through the crystal lattice. Some of these crystals are called uniaxial because there is only one direction (*optic axis*) along which their behavior is isotropic and double refraction of light does not take place.

In uniaxial crystals there are two refractive indices: one constant n_o, called ordinary, and the other n_e, variable (in most cases) with direction of propagation, called extraordinary. If light moves through the optic axis, where $n_o = n_e$, the behavior of the crystal is isotropic. In any other direction the light, moving through the crystal, is split into two beams of linearly polarized light mutually orthogonal. The ordinary ray always follows the Snell's law, the other only in few conditions. This feature is called birefringence.

In most cases the two beams (ordinary and extraordinary) follow different directions (Fig. 3.8) but in some cases they, always mutually orthogonal, have the same direction (examples a and b of Fig. 3.7) but different velocities. The most common uniaxial crystals are calcite and quartz; their refractive indices are, for $\lambda = 5893$ nm, $n_o = 1.6584$, $n_e = 1.4864$ for the first crystal and $n_o = 1.5443$, $n_e = 1.5534$ for the second one. The first is called a negative crystal ($n_e < n_o$) and the second positive ($n_e > n_o$). In Fig. 3.8 the subscripts 1 and 2 in r_1 and r_2 refer to ordinary and extraordinary rays for a negative crystal; for a positive crystal r_1 is the extraordinary ray and r_2 the ordinary. For a negative crystal extraordinary ray is then faster than the ordinary ray. For a positive crystal ordinary ray is faster than the extraordinary ray.

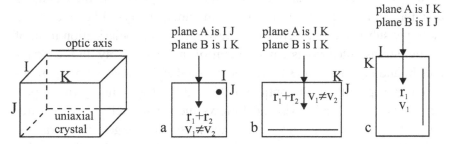

Fig. 3.7 In the three examples (a,b,c) the A plane of incidence is shown. The B plane is the boundary surface, between the air and the crystal where refraction happens, normal to to the plane A.

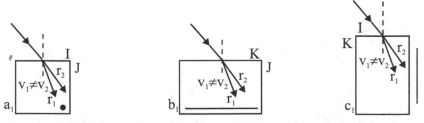

Fig. 3.8 In the three examples the A plane of incidence is shown. The B plane is the boundary surface, between the air and the crystal where refraction happens, normal to the plane A.

In the situations illustrated in Fig. 3.7 and Fig. 3.8 the plane of incidence A is the plane of the figures and the plane B, normal to A, is the boundary surface between of the air and the plane of the crystal, where the refraction happens.

The example b of Fig. 3.7 is the standard situation used for a retarder. The phase determined by a retarder is defined by the formula

$$\varphi = kt(n_o - n_e) \qquad k = \frac{2\pi}{\lambda}$$

and is positive for negative uniaxial crystals and for negative for positive uniaxial crystals.

Fig. 3.7 and Fig. 3.8 deserve more attention. The field E_o of the ordinary ray is parallel to the plane A in the examples a and a_1. The field E_o of the ordinary ray is normal to plane A in the examples b, b_1 and c_1. The field E_e of the extraordinary ray is parallel to the plane A in the example b. The field E_e of the extraordinary ray is normal to the plane A in the examples a and a_1. In the example c of Fig. 3.7 the uniaxial crystal behaves as an isotropic substance and there isn't birefringence: the polarization form of a single transmitted ray is equal to that of the incident ray. In the examples b_1 and c_1 of the Fig. 3.8 the electric field E_e vibrates neither perpendicular nor parallel to the optic axis (Fig. 3.9). The ordinary ray is perpendicular to the optic axis and follows the usual Snell's law of refraction; the extraordinary ray, changing direction changes, its refractive index (between n_o and n_e for negative crystals and n_e and n_o for positive crystals) and therefore it is difficult to find the angle of refraction of the extraordinary ray, given the angle of incidence.

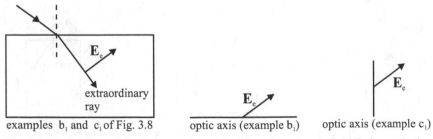

examples b_1 and c_1 of Fig. 3.8   optic axis (example b_1)   optic axis (example c_1)

Fig. 3.9 The left part of the figure can be misleading. The electric field E_e is not in the plane of incidence (the plane of the page)

Birefringence polarizers are fabricated assembling two or more prisms of uniaxial crystals cemented together with the Canada balsam or similar adhesive substance. The prisms never are in the condition of the examples b_1 and c_1 of the Fig. 3.8; however they are always in one of the conditions of examples a, b, c of the Fig. 3.7 and of the example a_1 of the Fig. 3.8. Only in these conditions the behavior of the extraordinary ray is easily predictable because this also follows the Snell's law.

As the refraction properties so the absorption changes in uniaxial crystals changing the direction of the plane of vibration of the electric field either the incident ray is normal to the surface of separation (plane B in Fig. 3.7) or oblique (plane B in Fig. 3.8). In the following problems we shall assume a zero value for the absorption.

3.1.5 Vectors for forms of polarization and matrices for polarizers and retarders

The main reference for the directions remains the plane xy with the z axis normal to this: for the polarizer θ is the angle between the x axis and the transmission axis of the polarizer (Fig. 3.1 where the line d_1-d_2 is the transmission axis); for retarder β is the angle between the x axis and the optic axis (Fig. 3.2). The angle φ is the phase difference, between the y component and the x component of the electric field when these pass through the retarder.

Representing with a plane harmonic wave the beam of a linearly polarized light, moving along the z axis and with the direction of the plane of oscillation of **E** lying in the plane xy, we can write, using complex notation, for the scalar value of the electric field

$$E = E_m e^{i\varepsilon} \qquad \varepsilon = kz - \omega t$$

and for its component

$$E_x = A_x \sin\varepsilon \quad E_y = A_y \sin\varepsilon \quad A_x = E_M \cos\theta \quad A_y = E_M \sin\theta$$

The intensity will be

$$I = A_x^{\,2} + A_y^{\,2} = E_M^{\,2}(\cos^2\theta + \sin^2\theta) = E_M^{\,2}$$

The general form for the intensity is, using the superscript character * for the conjugate value of a complex number,

$$I = \begin{vmatrix} A_x{}^* & A_y{}^* \end{vmatrix} \begin{vmatrix} A_x \\ A_y \end{vmatrix} = A_x{}^* A_x + A_y{}^* A_y$$

equal to the previous formula if A_x and A_y are real quantities.

When there is a phase difference φ between the x and y components we can write, using the equivalent matrix notation

$$\begin{vmatrix} E_x \\ E_y \end{vmatrix} = \begin{vmatrix} A_x \\ A_y e^{i\varphi} \end{vmatrix} e^{i\varepsilon}$$

The Jones vector

$$\begin{vmatrix} A_x \\ A_y e^{i\varphi} \end{vmatrix}$$

with its basic elements (the peak amplitudes of the x and y component and phase difference between them) has the necessary information for the description of any form of polarized light of a given intensity I.

In the following discussion and problems we shall assume that the incident beam on polarizers or retarders has a *unit intensity*, $E_M = 1$. Hence

$$A_x = \cos\theta \qquad A_y = \sin\theta$$

$$\begin{vmatrix} E_x \\ E_y \end{vmatrix} = \begin{vmatrix} \cos\theta \\ \sin\theta e^{i\varphi} \end{vmatrix} e^{i\varepsilon}$$

and, as expected

$$I = \begin{vmatrix} \cos\theta & \sin\theta e^{-i\varphi} \end{vmatrix} \begin{vmatrix} \cos\theta \\ \sin\theta e^{i\varphi} \end{vmatrix} = 1$$

When $\varphi = \pm\pi/2$ the elliptic form of polarization becomes

$$\frac{E_x{}^2}{A_x{}^2} + \frac{E_y{}^2}{A_y{}^2} = 1 \qquad A_x = E_M \cos\theta \qquad A_y = E_M \sin\theta$$

and with $E_M = 1$

$$\frac{E_x^{\,2}}{\cos\theta^2}+\frac{E_y^{\,2}}{\sin\theta^2}=E_M^{\,2}=I=1$$

Then the form of circular polarization ($\theta = 45°$) becomes

$$E_x^{\,2}+E_y^{\,2}=\frac{1}{2}$$

The major and minor semi-axes of the ellipse are

$$a=\cos\theta \qquad b=\sin\theta$$

and the ellipticity is defined as the ratio

$$\eta=\frac{b}{a}=\frac{\sin\theta}{\cos\theta}=\tan\theta$$

When the ellipticity is equal to one the form of polarization is circular. In fact

$$\theta=\arctan(1)=45°$$

The ellipse has the major axis along the x axis or along the y axis if the optic axis is parallel to the x or to the y axis. The major axis can subtend an angle β with the x axis if the optic axis subtends the same angle. The angle θ, on the contrary, define the values of the semi-axes.

The beam emerging from the quarter-wave retarder has a linear form of polarization if direction of oscillation of the electric field and that of the optic axis coincide. For example, if direction of the oscillation of the electric field and that of the optic axis are both at 0° or 90° from the x axis the beam emerging from the retarder has the same form of linear polarization of the beam entering the retarder. When these directions are different the form of polarization of the emerging beam is non linear.

The phase difference, with our convention, is always given by

$$\varphi=\varphi_y-\varphi_x$$

When φ_x or φ_y are calculated we have to remember that

$$i = e^{i\frac{\pi}{2}} = i\sin(\frac{\pi}{2}) \qquad -i = e^{-i\frac{\pi}{2}} = i\sin(-\frac{\pi}{2})$$

The imaginary unit i implies a phase of 90° or -90° (see Sec. 3.2.4). Properly speaking the *phase shift* would be defined as $\sin(\varphi)$ remembering its first appearance in a sine function. The general Jones vector is

$$\begin{vmatrix} \cos\theta \\ \sin\theta e^{i\varphi} \end{vmatrix}$$

For example, a linear polarized light ($\varphi = 0$) of *unit intensity* with direction of oscillation along the x axis or subtending an angle $\theta = 0°$, $\theta = 45°$, $\theta = 90°$ $\theta = 135°$ or $\theta = 180°$ will have the following Jones vectors

$$\begin{vmatrix} 1 \\ 0 \end{vmatrix} \quad \frac{1}{\sqrt{2}}\begin{vmatrix} 1 \\ 1 \end{vmatrix} \quad \begin{vmatrix} 0 \\ 1 \end{vmatrix} \quad \frac{1}{\sqrt{2}}\begin{vmatrix} -1 \\ 1 \end{vmatrix} \quad \begin{vmatrix} -1 \\ 0 \end{vmatrix}$$

In a similar way can be determined the vectors when φ is equal to π or to $\pi/2$ for the corresponding form of polarized light, always of *unit intensity*, emerging from a retarder.

The form of polarization that polarizers and retarders produce into a beam passing through them, is represented by a two by two *Jones matrix*

$$m = \begin{vmatrix} a & b \\ c & d \end{vmatrix}$$

If a beam of light, whose Jones vector for its form of polarization is

$$v_1 = \begin{vmatrix} k \\ h \end{vmatrix}$$

passes through a device whose matrix is m_1 the form of polarization v_2 of the light emerging from the device will be

$$v_2 = m_1 v_1$$

If in series there is another device whose matrix is m_2 the form of polarization v_3 of the light emerging from the second device will be

$$v_3 = m_2 v_2 = m_2 m_1 v_1$$

If the devices in series are n, the final form v_f of polarization emerging from the last device will be

$$m_n\, m_{n-1}\,...m_2\, m_1 v_1 = v_f$$

As a rule in the following problems, *we will not explicitly indicate the product v*v of the vector v by its transpose v** (that is also its conjugate, if the vector is complex) when we are going to calculate the intensity. The calculus of this product of matrices is an easy task using MATLAB. A list of some commonly used Jones matrices follows.

linear polarizer

$$(\theta = 0)\begin{bmatrix} 1 & 0 \\ 0 & 0 \end{bmatrix} \qquad (\theta = \pm\frac{\pi}{4})\ \frac{1}{2}\begin{bmatrix} 1 & \pm 1 \\ \pm 1 & 1 \end{bmatrix} \qquad (\theta = \frac{\pi}{2})\begin{bmatrix} 0 & 0 \\ 0 & 1 \end{bmatrix}$$

and for any value of θ

$$\begin{vmatrix} C_0^2 & C_0 S_0 \\ C_0 S_0 & S_0^2 \end{vmatrix} \qquad C_0 = \cos\theta \qquad S_0 = \sin\theta$$

quarter-wave linear retarders $(\varphi = \pi/2)$

$$(\beta = 0)\begin{bmatrix} 1 & 0 \\ 0 & -i \end{bmatrix} \qquad (\beta = \pm\frac{\pi}{4})\ \frac{1}{\sqrt{2}}\begin{bmatrix} 1 & \pm i \\ \pm i & 1 \end{bmatrix} \qquad (\beta = \frac{\pi}{2})\begin{bmatrix} -i & 0 \\ 0 & 1 \end{bmatrix}$$

$$(\beta\text{:any value})\begin{bmatrix} C_1^2 - iS_1^2 & C_1 S_1(1+i) \\ C_1 S_1(1+i) & -iC_1^2 + S_1^2 \end{bmatrix} \qquad C_1 = \cos\beta \quad S_1 = \sin\beta$$

half-wave retarders $(\varphi = \pi)$

$$(\theta = 0)\begin{bmatrix} 1 & 0 \\ 0 & -1 \end{bmatrix} \qquad (\theta = \pm\frac{\pi}{4})\begin{bmatrix} 0 & \pm 1 \\ \pm 1 & 0 \end{bmatrix} \qquad (\theta = \frac{\pi}{2})\begin{bmatrix} -1 & 0 \\ 0 & 1 \end{bmatrix}$$

and for any value of β

$$\begin{vmatrix} C_2 & S_2 \\ S_2 & -C_2 \end{vmatrix} \qquad C_2 = \cos 2\beta \qquad S_2 = \sin 2\beta$$

3.2 Problems

3.2.1 Natural light

A beam R_1 of monochromatic natural light of intensity I_1 is incident on the upper plane surface of a sheet of a transparent substance whose refractive index is $n = 1.6$. The beam R_1 is moving in the air. The reflected ray R'_1 is polarized (Fig. 3.10). The refracted ray R_2 is again reflected (ray R'_2) in the medium and refracted in the air (R_3). Assuming there isn't absorption from the transparent substance find reflectivity and refractivity when the light is reflected and refracted on the upper surface.

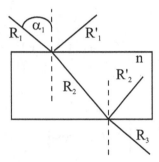

Fig. 3.10

Find the reflexivity of the rays R'_1, R'_2 and refractivity of the rays R_2, R_3, the degree of polarization of R'_1 and R_2 and the form of polarization of the rays R'_1, R_2, R'_2, and R_3.

Solution:

The electric field of the reflected ray is oscillating only in the plane normal to the plane of incidence and the incident ray subject to the Brewster condition (Fig. 3.11), has the angle of incidence

$$\alpha_1 = \arctan n = 58.0°$$

Hence

$$\alpha_2 = \arcsin\left(\frac{\sin \alpha_1}{n}\right) = 32.0°$$

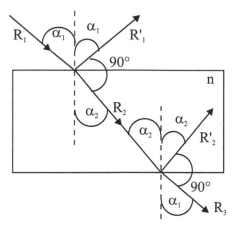

Fig. 3.11

Since the direction of vibration of the electric field (and the corresponding angle γ defined in Sec. 3.1.3) of R_1 varies rapidly in random manner
and mean value of $\sin^2 \gamma$ and $\cos^2 \gamma$ is 1/2 we can write for the intensity
corresponding to the component of the electric field normal to the plane
of incidence (using notations of Sec. 3.1.3)

$$I_{1y} = \frac{1}{2} I_1$$

and then for reflectivity

$$R = \frac{I_{ry}}{I_1} = I_{rw} \cos^2 \gamma + I_{ry} \sin^2 \gamma = \frac{1}{2} I_{ry} = \frac{1}{2} \frac{\sin^2(\alpha_1 - \alpha_2)}{\sin^2(\alpha_1 + \alpha_2)} = 0.1$$

remembering that $I_{rw} = 0$, $\gamma = 45°$ and with the value of I_{ry} (the ratio of
square sine) equal to 0.2.
The refractivity will be directly (by the principle of energy conservation)

$$T = \frac{I_2}{I_1} = \frac{I_1 - 0.1 I_1}{I_1} = 0.9$$

or using the formulae

$$I_{tw} = \frac{\sin 2\alpha_1 \sin 2\alpha_2}{\sin^2(\alpha_1 + \alpha_2)\cos^2(\alpha_1 - \alpha_2)} = 1$$

$$I_{ty} = \frac{\sin 2\alpha_1 \sin 2\alpha_2}{\sin^2(\alpha_1 + \alpha_2)} = 0.8$$

$$T = \frac{1}{2}(I_{tw} + I_{ty}) = \frac{1}{2}(1 + 0.8) = 0.9$$

Intensities I_{2w} and I_{2y} will be directly

$$I_{2w} = 0.5 I_1 \qquad I_{2y} = I_{1y} - I_{ry} = (0.5 - 0.1)I_1 = 0.4 I_1$$

finding again

$$I_{tw} = \frac{I_{2w}}{I_{1w}} = \frac{0.5 I_1}{0.5 I_1} = 1 \qquad I_{ty} = \frac{I_{2y}}{I_{1y}} = \frac{0.4 I_1}{0.5 I_1} = 0.8$$

The degrees of polarization for R'$_1$ and R$_2$ are

$$P_r = \left| \frac{I_{rw} - I_{ry}}{I_{rw} + I_{ry}} \right| = 1 \qquad P_t = \left| \frac{I_{tw} - I_{ty}}{I_{tw} + I_{ty}} \right| = 0.1$$

The reflected ray R'$_1$ is linearly polarized and so P_r is equal to one, as it is expected, but the refracted ray R$_2$ is only partially polarized. The angles of reflection and refraction on the lower surface are now

$$\alpha''_1 = \alpha_2 = 32° \qquad \alpha''_2 = \alpha_1 = 58°$$

Again we are in the Brewster condition because

$$\alpha''_1 + \alpha''_2 = 90°$$

So it will be

$$R'' = I''_{ry} = I_{ty} = 0.8$$

because the reflected ray R'$_2$ is linearly polarized. The refractivity is

$$T'' = T - I''_{ry} = T - I_{ty} = 0.1$$

and the refracted ray R_3 is only partially polarized.

According to the principle if conservation of energy we have

$$R+R''+T''=1$$

80% of the energy is associated with the ray R'_2 and 10% with R'_1 and R_3 respectively.

3.2.2 Linearly polarized light

A beam R_1 of monochromatic linearly polarized light of intensity I_1 is incident, with an angle α_1 on the upper plane surface **B** of a sheet of a transparent substance whose refractive index is $n = 1.5$. The beam R_1 is moving in the air. The plane of oscillation of the electric field of the ray R_1 subtends an angle γ with the plane of incidence **A**. Assuming there isn't absorption from the transparent substance find the relative intensities due to the components of the intensities on the surfaces **A** and **B** for the reflected and refracted rays, the reflectivity and refractivity, the degrees of polarization for the reflected and refracted rays when, for the fixed values 0° and 90° of γ, α_1 assumes the values 40°, Brewster angle and 70°.

Solution:

In the Brewster condition we have

$$\alpha_1 = \alpha_B = \arctan(n) = 56.3° \qquad \alpha_2 = 33.7° \qquad \alpha_1 + \alpha_2 = 90°$$

The intensities of the components of the electric field

$$I_{rw} = \frac{\tan^2(\alpha_1 - \alpha_2)}{\tan^2(\alpha_1 + \alpha_2)}$$

$$I_{ry} = \frac{\sin^2(\alpha_1 - \alpha_2)}{\sin^2(\alpha_1 + \alpha_2)}$$

$$I_{tw} = \frac{\sin 2\alpha_1 \sin 2\alpha_2}{\sin^2(\alpha_1 + \alpha_2)\cos^2(\alpha_1 - \alpha_2)}$$

$$I_{ty} = \frac{\sin 2\alpha_1 \sin 2\alpha_2}{\sin^2(\alpha_1 + \alpha_2)}$$

depend from α_1 and n. They are independent from γ.
Refractivity and reflectivity depend from α_1, n and from γ, too

$$R = I_{rw} \cos^2 \gamma + I_{ry} \sin^2 \gamma$$

$$T = I_{tw} \cos^2 \gamma + I_{ty} \sin^2 \gamma$$

With the assigned values of α_1 and of γ the requested quantities are in
Table 3.1. Increasing α_1 R increases and T decreases. $\gamma = 0°$ implies $E_{1y} = 0$: so in the Brewster condition there isn't reflection and there is $R=0$, $T=1$
and $P_r = 1$. The values of the relative intensities lying in the plane w of
incidence and in the plane y of the boundary surface satisfy the principle
of conservation of the energy because we have: $I_{rw} + I_{tw} = I_{ry} + I_{ty} = 1$.

Table 3.1

$\gamma°$	$\alpha_1°$	$\alpha_2°$	I_{rw}	I_{ry}	I_{tw}	I_{ty}	R	T	P_r	P_t
0	40	25.4	0.01	0.08	0.99	0.92	0.01	0.99	0.7	0.3
0	56.3	33.7	0	0.1	1	0.9	0	1	1	0.01
0	70	38.8	0.04	0.3	0.96	0.7	0.04	0.96	0.75	0.15
90	40	25.4	0.01	0.08	0.99	0.92	0.08	0.92	0.69	0.03
90	56.3	33.7	0	0.15	1		0.85	0.15	0.85	1	0.08
90	70	38.8	0.04	0.30	0.96	0.70	0.3	0.7	0.75	0.15

The degree of polarization for the reflected ray

$$P_r = \left| \frac{I_{rw} - I_{ry}}{I_{rw} + I_{ry}} \right|$$

has the value one for $\alpha_1 = \alpha_{1B}$ and, of course independent from γ.
The degree of polarization for the refracted ray is never equal to one.

$$P_t = \left| \frac{I_{tw} - I_{ty}}{I_{tw} + I_{ty}} \right|$$

Refractivity is greater than reflectivity, for fixed low values of α_1, varying γ from 0° to 90° (in Fig. 3.12 are given refractivity and reflectivity for the fixed value 70° of α_1). On the contrary refractivity is lower than reflectivity, for fixed high values of α_1, varying γ from 0° to 90°.

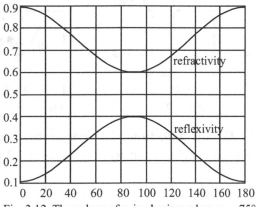

Fig. 3.12 The values of γ in abscissa when $\alpha_1 = 75°$

3.2.3 Half-wave retarder 1

A beam of a monochromatic ($\lambda = 0.589\ \mu$) and linearly polarized light, of intensity I_1 and moving in the air along the z axis, reaches a half-wave quartz retarder placed in the xy plane. The plane where the electric field E_1 of the beam is vibrating coincides with the x axis. The optics axis of the retarder subtends an angle $\theta = 60°$ with the x axis. After the retarder is placed, parallel, a polarizer whose transmission axis subtends also an angle $\theta = 60°$ with the x axis. Find the angle that the plane of oscillation of the electric field E_1 will subtend with the x axis when the beam emerges from the retarder and the intensity I_2 of the beam emerging from the polarizer. If the thickness t (length along the z axis) is about one millimeter find its exact value within two correct decimal digits. Find also the m value by which $\lambda/2$ must be multiplied to obtain the requested value of t. The refractive indices of the quartz, with the assigned value of λ are $n_o = 1.5443$ and $n_e = 1.5534$. Assume that polarizer doesn't modify the intensity of the light.

Solution:

The half-wave plate changes the angle 60° between E_1 and the optic axis into the angle -60° from the optic axis or 120 from the x axis (Fig. 3.13).

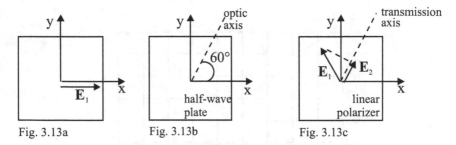

Fig. 3.13a   Fig. 3.13b   Fig. 3.13c

The electric field E_1 emerging from the retarder becomes E_2 when it emerges from polarizer with an intensity I_2 given by the Malus' law

$$I_2 = I_1 \cos^2 60° = (0.5)^2 I_1) = \frac{I_1}{4}$$

The values of t and m satisfying our conditions are found plotting their linear function

$$t = \frac{\lambda}{2(n_e - n_o)} m$$

After a short "trial and error" procedure we find (Fig. 3.14)

$$m = 31 \qquad d = 1.0032 = 1.00 \, \text{mm}$$

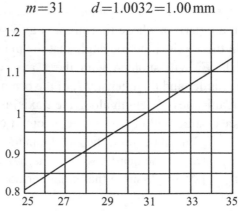

Fig. 3.14 Values of t (mm) function of m

3.2.4 Retarders

A beam of natural monochromatic ($\lambda = 0.5893\ \mu$) light, of the intensity I_1, is moving along the z axis and reaches a polarizer P_1, whose plane is coincident with xy plane. The transmission axis of P_1 coincides with the y axis (Fig. 3.15a). The beam emerging from P_1 reaches a retarder that has its plane parallel to the plane xy. The optic axis subtends an angle $\beta = 45°$ with the x axis (Fig. 3.15b). The beam emerging from the retarder reaches another polarizer P_2, placed behind the retarder, whose transmission axis has the same direction of the first one (Fig. 3.15b). Find the intensity of the beam emerging from the polarizer P_2 when the retarder is half-wave R_{h-w} or quarter-wave R_{q-w}. If another half-wave plate P_{h-w} (whose optic axis is at 45° from the x axis) takes the place of the polarizer P_2, behind the half-wave R_{h-w} or the quarter-wave R_{q-w} (Fig. 3.16), find the polarization form of the final beams and their intensities. Assume that polarizers and retarders don't absorb light.

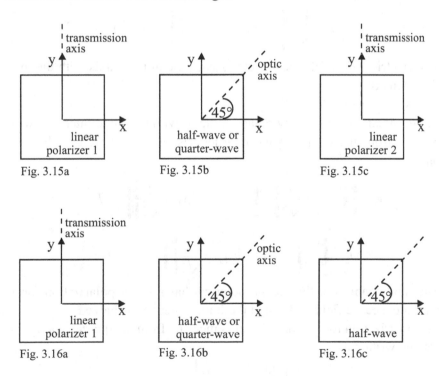

Fig. 3.15a   Fig. 3.15b   Fig. 3.15c

Fig. 3.16a   Fig. 3.16b   Fig. 3.16c

Solution:

a. In series are P_1, $R_{h\text{-}w}$, P_2:

If we have a half-wave retarder, the electric field $\mathbf{E}_1$ is rotated clockwise of $90°$ from the y to the x axis: no beam will emerge from the second polarizer and the intensity of light becomes zero. Using the matrix notation we have

$$v_1 = \begin{vmatrix} 0 \\ 1 \end{vmatrix} \quad m_1 = \begin{vmatrix} 0 & 1 \\ 1 & 0 \end{vmatrix} \quad v_r = m_1 v_1 = \begin{vmatrix} 1 \\ 0 \end{vmatrix} \quad I_r = 1$$

$$m_2 = \begin{vmatrix} 0 & 0 \\ 0 & 1 \end{vmatrix} \quad v_2 = m_2 v_r = \begin{vmatrix} 0 \\ 0 \end{vmatrix} \quad I_2 = 0$$

b. In series are P_1, $R_{h\text{-}w}$, $P_{h\text{-}w}$:

If a half-wave plate $P_{h\text{-}w}$ takes the place of the polarizer P_2, behind the half-wave $R_{h\text{-}w}$ we have

$$m_2 = \begin{vmatrix} 0 & 1 \\ 1 & 0 \end{vmatrix} \quad v_2 = m_2 v_r = \begin{vmatrix} 0 \\ 1 \end{vmatrix} \quad I_2 = 1$$

The electric field of the beam rotates back along the y axis with the initial intensity.

c. In series are P_1, $R_{q\text{-}w}$, P_2:

In this situation a circularly polarized light, with the same intensity, will emerge from the quarter-wave retarder. Using the matrix calculus we have

$$v_1 = \begin{vmatrix} 0 \\ 1 \end{vmatrix} \quad m_1 = \frac{1}{\sqrt{2}} \begin{vmatrix} 1 & i \\ i & 1 \end{vmatrix} \quad v_r = m_1 v_1 = \frac{1}{\sqrt{2}} \begin{vmatrix} i \\ 1 \end{vmatrix} \quad I_r = 1$$

$$m_2 = \begin{vmatrix} 0 & 0 \\ 0 & 1 \end{vmatrix} \quad v_2 = m_2 v_r = \frac{1}{\sqrt{2}} \begin{vmatrix} 0 \\ 1 \end{vmatrix} \quad I_2 = 0.5$$

The final beam after the second polarizer has a linear polarization form with the electric field at $45°$ and the initial intensity is halved.

The polarization form v_r of the beam emerging from the quarter-wave can also be written as

$$v_r = \frac{1}{\sqrt{2}} \begin{vmatrix} i \\ 1 \end{vmatrix} = \begin{vmatrix} \cos\theta\, e^{i\frac{\pi}{2}} \\ \sin\theta \end{vmatrix} \qquad \theta = 45°$$

The ellipticity and phase difference are

$$\eta = 1 \qquad \varphi = \varphi_y - \varphi_x = -\frac{\pi}{2}$$

So the polarization form is circular and handedness is left (see Sec. 3.1.2).

d. In series are P_1, $R_{q\text{-}w}$, $P_{h\text{-}w}$:

If a half-wave plate takes the place of the second polarizer we have

$$v_1 = \begin{vmatrix} 0 \\ 1 \end{vmatrix} \quad m_1 = \frac{1}{\sqrt{2}}\begin{vmatrix} 1 & i \\ i & 1 \end{vmatrix} \quad v_r = m_1 v_1 = \frac{1}{\sqrt{2}}\begin{vmatrix} i \\ 1 \end{vmatrix} \quad I_r = 1$$

$$m_2 = \begin{vmatrix} 0 & 1 \\ 1 & 0 \end{vmatrix} \quad v_2 = m_2 v_r = \frac{1}{\sqrt{2}}\begin{vmatrix} 1 \\ i \end{vmatrix} = \begin{vmatrix} \cos\theta \\ \sin\theta\, e^{i\frac{\pi}{2}} \end{vmatrix} \quad \theta = 45° \quad I_2 = 1$$

The polarization form of the final beam, emerging from the second half-wave retarder, is now circular ($\eta = 1$), the handedness is reversed to the right ($\varphi = \pi/2$) and the intensity is now $I_2 = 1$.

3.2.5 A calcite plate

A beam of a monochromatic ($\lambda = 0.5893\ \mu$) natural light of intensity I_1 is incident normally on a polarizer and with an angle $\alpha = 45°$ on the plane boundary surface of a calcite crystal that has $h = 1$ cm, $n_o = 1.6584$, and $n_e = 1.4864$ (Fig. 3.17). The direction of the transmission axis of the polarizer can rotate from a direction normal to the plane of incidence (that

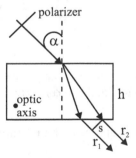

Fig. 3.17

of the page) to a direction parallel to this plane. We shall call θ the angle between these directions varying from 0° to 90°. Indicate for the values of θ in the interval $(0, 90)°$ if the emerging rays from the crystal are always two and which of r_1 and r_2 is the ordinary and extraordinary ray. Whel both rays emerge parallel, from the lower surface of the calcite plate, find the distance s between them. Find the values of intensity for the ordinary and extraordinary rays varying θ in the interval $(0, 90)°$ assuming that there isn't different absorption due to the different orientation of the electric field (see Sec. 3.1.4).

Solution:

If polarizer has a D_N direction the incident beam will have the electric field E_1 parallel to the optic axis and only extraordinary ray emerges from the lower surface of the crystal; if polarizer has a D_p direction the incident beam will have the electric field normal to the optic axis and only ordinary ray emerges. For the other directions of the transmission axis both rays will emerge (Fig. 3.18) and the angles of refracted rays are

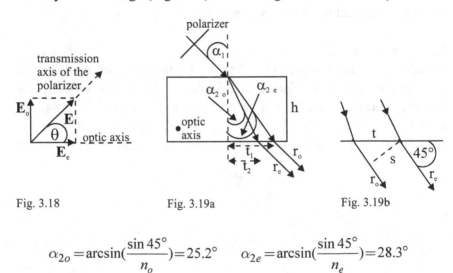

Fig. 3.18   Fig. 3.19a   Fig. 3.19b

$$\alpha_{2o} = \arcsin(\frac{\sin 45°}{n_o}) = 25.2° \qquad \alpha_{2e} = \arcsin(\frac{\sin 45°}{n_e}) = 28.3°$$

Hence r_1 is the ordinary ray and r_2 is the extraordinary (Fig. 3.19a). The distance s is (Fig. 3.19b)

$$t = t_2 - t_1 = h(\tan\alpha_{2e} - \tan\alpha_{2o}) = 0.68\,\text{mm} \qquad s = t\sin 45° = 0.5\,\text{mm}$$

The electric fields of the ordinary and extraordinary rays and the corresponding intensities are

$$E_e = E_1 \cos\theta \quad I_e = I_1 \cos^2\theta \qquad E_o = E_1 \sin\theta \quad I_o = I_1 \sin^2\theta$$

These formulae (Fig. 3.20) are in agreement with the previous result regarding the emerging of one or two rays from the lower surface of the crystal.

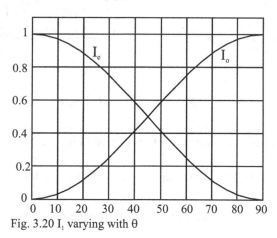

Fig. 3.20 I_1 varying with θ

3.2.6 A calcite prism

A beam of monochromatic ($\lambda = 0.5893 \ \mu$) and linearly polarized light moving along the z axis, reaches normally the surface AB of a prism of calcite ($n_o = 1.6584$ and n_e 1.4864). The direction of oscillation of the electric field subtends angle θ with the plane of incidence. The optic axis is parallel to the surface AB (Fig. 3.21). Find what happens to the beam through the prism and when it emerges from the surface BC.

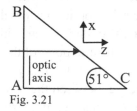

Fig. 3.21

Solution:

The electric field of the incident beam can be considered (Fig. 3.22a) in its component normal and parallel to the optic axis

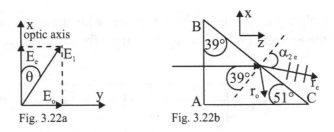

Fig. 3.22a   Fig. 3.22b

$$E_o = E_1 \sin\theta \qquad E_e = E_1 \cos\theta$$

For the rays incident on surface BC we have (Fig. 3.22b)

$$\sin n\alpha_{2o} = n_o \sin 39° = 1.04 \quad \sin\alpha_{2e} = n_e \sin 39° = 0.93 \quad \alpha_{2e} = 68.4°$$

We have $\sin\alpha_{2o} > 1$ and the critical angle for the ordinary ray

$$\alpha_{1o} = \arcsin(\frac{1}{n_o}) = 37.08°$$

Hence the ordinary ray, whose electric field oscillates in the *yz* plane is totally reflected, whereas the extraordinary ray, whose electric field oscillates in the *xz* plane, is refracted emerging, from the surface BC, with the angle 68.4° (Fig. 3.22b).

3.2.7 A quartz prism

A quartz prism ($n_o = 1.5443$, $n_e = 1.5534$), whose section is a triangle with AB = 1 cm and BC = 3 cm has optic axis parallel to BC (Fig. 3.23). A beam of monochromatic ($\lambda = 0.5893$ μ) and linearly polarized light, of intensity *I*, reaches normally the mid point of the surface BC. The electric field *E* oscillates in the plane *xy* subtending an angle $\theta = 45°$ with the *x* axis. Find the phase difference, between the waves associated to the ordinary and extraordinary rays, produced after traveling from D to D', the angles of refraction of both rays emerging from the surface AC.

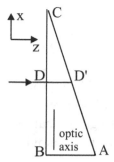

Fig.. 3.23

Solution:

Moving through the prism ABC the beam is divided into two rays, even if superimposed: one, the extraordinary, has the electric field oscillating in the plane *xz*; the other, ordinary, has the electric field oscillating in the plane *yz*. The path DD' = *t* (Fig. 3.24a) is

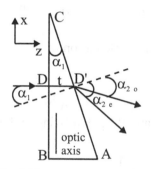

Fig. 3.24a

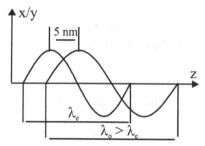

Fig. 3.24b λ_o must be seen in the plane yz while λ_e is in the plane xz

$$t = \frac{BC}{2} \sin \alpha_1 = 0.47 \, \text{cm}$$

where α_1 is the angle between the sides CB and CA

$$\alpha_1 = \arctan(\frac{BA}{BC}) = \arctan(\frac{1}{3}) = 18.4°$$

The two rays reach the surface AC with different phases

$$\varphi_o = \frac{2\pi}{\lambda} n_o t = 87.50° \qquad \varphi_e = \frac{2\pi}{\lambda} n_e t = 90.39°$$

Their phase difference is

$$\Delta\varphi=\varphi_e-\varphi_o=2.9°=0.05\,\text{rad}$$

corresponding to a distance of 0.005 µ (or 5 nm) between the peaks (Fig. 3.24b). It is known that to a lower refractive index correspond a faster velocity and a longer wavelength so that the period T and the frequency v remain constant if the same ray is moving in substances with two different refractive indices.

The angle α_1 between the incident rays on AC and the normal to this side is equal to the angle between the sides CB and CA (indeed they are both complement of the angle between the sides CD' and DD'). Then for the refracted rays emerging in the air, subtending the angles α_{2o} and α_{2e} with the normal to the surface CA, we have (Fig. 3.23a)

$$n_o\sin\alpha=\sin\alpha_{2o}\quad \alpha_{2o}=29.23°\qquad n_e\sin\alpha=\sin\alpha_{2e}\quad \alpha_{2e}=29.44°$$

The refracted rays maintain in the air the same orientation of the electric field that they had in the crystal.

3.2.8 Two calcite prisms 1
A beam of monochromatic ($\lambda=0.5893$ µ) natural light, moving in the air,

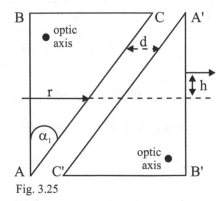

Fig. 3.25

is incident normally on the face AB of the first of two prisms of calcite ($n_o=1.6584$ and n_e 1.4864), spaced d = 1 mm apart and with the directions of the optic axes as shown in Fig. 3.25. The wedge angle is $\alpha_1 =$ 40°. Define the directions of the rays, and of the oscillations of the associated electric fields, crossing the first prism, the air gap, the second prism and emerging from surface A'B'.

Solution:

The critical angles for ordinary and extraordinary rays on the boundary surface AC between calcite and air are

$$\alpha_{1oc} = \arcsin(\frac{1}{n_o}) = 37.1° \qquad \alpha_{1ec} = \arcsin(\frac{1}{n_e}) = 42.3°$$

Hence the ordinary incident ray r_o is entirely reflected. Whereas the incident extraordinary ray is refracted with angle (Fig. 3.26)

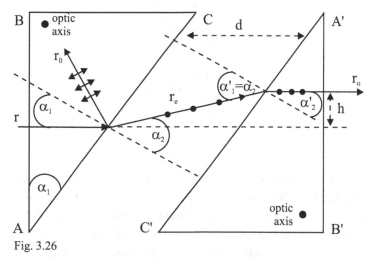

Fig. 3.26

$$n_e \sin\alpha_1 = \sin\alpha_2 \qquad \alpha_2 = \arcsin(n_e \sin\alpha_1) = 72.8°$$

The angle α'_1 of incident extraordinary ray on the surface A'C' is equal to α_2. The incident extraordinary ray moving in the air remains extraordinary when it is again refracted with the angle

$$\sin\alpha'_1 = n_e \sin\alpha'_2 \qquad \alpha'_2 = \arcsin(\frac{1}{n_e}\sin\alpha'_1) = 40°$$

Then the refracted ray emerges normal to the surface A'B'.

3.2.9 Two calcite prisms 2

A beam of monochromatic ($\lambda = 0.5893$ μ) natural light, moving in the air, is incident normally on the face AB of the first of two prisms of calcite ($n_o = 1.6584$ and $n_e = 1.4864$), cemented together and with the directions of the optic axes as shown in Fig. 3.27. Define the directions of the rays, and of the oscillations of the associated electric fields, crossing the prisms along the section AC and emerging from surface CD. Find the value of the wedge angle α_{1max} allowing the maximum deviation δ_{1max} between the two rays refracted from the surface CD on assumption that both are subject to refraction. Can we have a greater deviation if two prisms of quartz ($n_o = 1.5443$, $n_e = 1.5534$) are used?

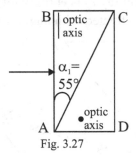

Fig. 3.27

Solution:

The ordinary and extraordinary rays (Fig. 3.28) are superimposed when they move through the first prism with the electric field E_o oscillating in the plane xy and E_e in the plane xz. Moving through the second prism the ordinary ray (E_o) becomes extraordinary (E'_e) and conversely the extraordinary (E_e) becomes ordinary (E'_o). For the ordinary and extraordinary rays crossing the surface AC, remembering that from geometry the angle of incidence is equal to the wedge angle $\alpha_1 = 55°$, we have

$$n_e \sin \alpha_1 = n_o \sin \alpha_{2o} \qquad \alpha_{2o} = \arcsin\left(\frac{n_e}{n_o} \sin \alpha_1\right) = 47.2°$$

$$n_o \sin \alpha_1 = n_e \sin \alpha_{2e} \qquad \alpha_{2e} = \arcsin\left(\frac{n_o}{n_e} \sin \alpha_1\right) = 66.1°$$

The ordinary ray crossing the surface AC, where it becomes extraordi-

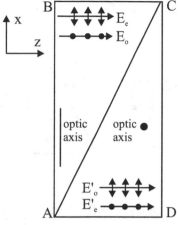

Fig. 3.28

nary, has an angle of refraction greater than $\alpha_1 = 55°$ because $n_e < n_o$. The ordinary ray is subject to refraction because its angle of incidence is lower of its critical angle

$$\alpha_1 < \alpha_{1c} = \arcsin(\frac{n_e}{n_o}) = 63.7°$$

The angles of incidence on the surface CD are

$$\alpha'_{1o} = \alpha_1 - \alpha_{2o} = 7.8° \qquad \alpha'_{1e} = \alpha_1 - \alpha_{2e} = 11.1°$$

and the corresponding angles of refraction for the rays emerging in the air

$$n_o \sin \alpha'_{1o} = \sin \alpha'_{2o} \qquad \alpha'_{2o} = \arcsin(n_o \sin \alpha'_{1o}) = 12.9°$$

$$n_e \sin \alpha'_{1e} = \sin \alpha'_{2e} \qquad \alpha'_{2e} = \arcsin(n_e \sin \alpha'_{1e}) = 16.6°$$

In Fig. 3.29 is shown only the ray tracing of the extraordinary ray in the first prism that becomes ordinary in the second and is incident and refracted on the surface CD.
Doing again the previous steps assuming that the prisms are of quartz we find that the emerging rays are practically superimposed and normal to the surface CD. Their agles are now

$$\alpha'_{2o}=0.74° \qquad \alpha'_{2e}=0.70°$$

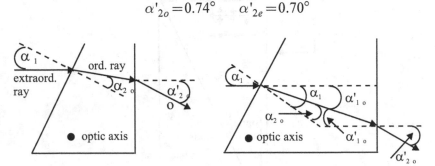

Fig. 3.29

Plotting (Fig. 3.31) the angles of refraction α'_{2e} versus the values of the incident angle $\alpha_1 < \alpha_{1c}$ for the ordinary ray on the surface AC, defined in the interval $(0, 63.6)°$ we find, with a wedge angle $(63.6°)$ a bit less than $63.7°$, $\alpha'_{2e} = 37.3°$ and $\alpha'_{2o} = 17.1°$. Hence the maximum deviation δ_{1max} between the two refracted rays is $37.3° + 17.1° = 54.4°$ (Fig. 3.30).

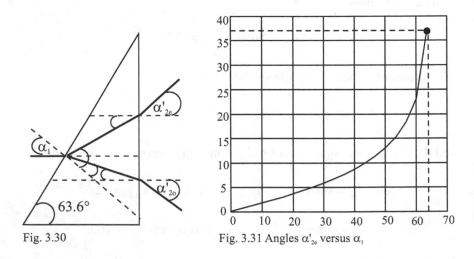

Fig. 3.30

Fig. 3.31 Angles α'_{2e} versus α_1

3.2.10 Two polarizers

A beam of natural light, having intensity $I_1 = 1$ and moving along the z axis is incident on a first polarizer P_1, lying in the xy plane and whose pass-plane is parallel to the x axis (Fig. 3.32). The beam emerging from

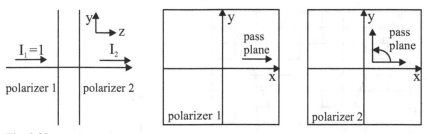

Fig. 3.32

P_1 is incident on a second parallel polarizer P_2, rotating around the z axis. Give the plot of the intensity I_2 emerging from P_2 when the angle θ between its pass-plane and the x axis is increasing from 0 to 180°. The polarizers, considered "ideal", don't absorb light. Use the matrix calculus.

Solution:
The beam emerging from the first polarizer doesn't change the intensity and reaches P_2 having the polarization form

$$v_1 = \begin{vmatrix} 1 \\ 0 \end{vmatrix}$$

The transmission axis of second polarizer, rotating around the z axis, subtends an angle θ variable with the x axis. Thereby P_2 has the matrix

$$m_1 = \begin{vmatrix} C_1^2 & C_1 S_1 \\ C_1 S_1 & S_1^2 \end{vmatrix} \qquad C_1 = \cos\theta \quad \cdot \quad S_1 = \sin\theta$$

The beam emerging from P_2 will have a variable form of linear polarization v_2 given by the product

$$v_2 = m_1 v_1$$

The corresponding variable intensities I_2 are hence easily calculated. The values of I_2 follow, as expected, the Malus' law (Fig. 3.33).

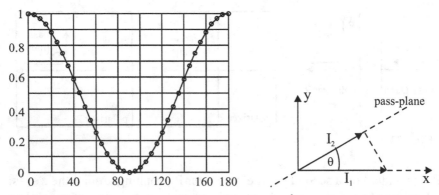

Fig. 3.33 Circles (o) are the intensities I_2 and the line $(\cos^2\theta)$ represents the Malus' law

3.2.11 Three polarizers

A beam of natural light, having intensity $I_1 = 1$ and moving along the z axis is incident on a first polarizer P_1, lying in the xy plane and whose pass-plane is parallel to the x axis (Fig. 3.34). The beam emerging from

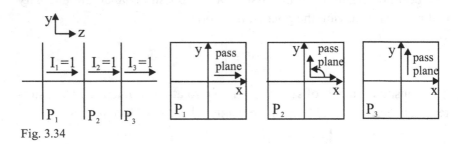

Fig. 3.34

P_1 is incident on a second polarizer P_2, parallel to first but rotating around the z axis. The beam emerging from P_2 is incident on a third polarizer P_3 whose pass-plane is normal to the x axis. Give the plot of the intensity I_2 emerging from P_2 when the angle θ between its pass-plane and the x axis is increasing from 0 to 360° and the intensity I_3 of the beam emerging from the third polarizer. The polarizers, considered "ideal", don't absorb light. Use the matrix calculus.

Solution:

The beam emerging from the first polarizer doesn't change the intensity and reaches P_2 having the polarization form

$$v_1 = \begin{vmatrix} 1 \\ 0 \end{vmatrix}$$

The transmission axis of second polarizer, rotating on the z axis, subtends an angle θ variable with the x axis. Thereby P_2 has the matrix

$$m_1 = \begin{vmatrix} C_1^2 & C_1 S_1 \\ C_1 S_1 & S_1^2 \end{vmatrix} \qquad C_1 = \cos\theta \qquad S_1 = \sin\theta \qquad \theta \text{ varying from 0 to } 2\pi$$

The beam emerging from P_2 will have a variable form of linear polarization v_2 given by the product

$$v_2 = m_1 v_1$$

and a corresponding variable intensity I_2 that follows, as expected, the Malus'law (Fig. 3.36).

The polarizer P_3 has the matrix

$$m_2 = \begin{vmatrix} 0 & 0 \\ 0 & 1 \end{vmatrix}$$

The beam emerging from P_3 has the polarization form

$$v_3 = m_2 m_1 v_1$$

and a corresponding variable intensity I_3 that follows, as expected, the Malus'law.

In fact we have (Fig. 3.35)

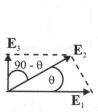

Fig. 3.35

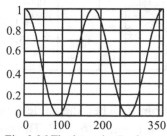

Fig. 3.36 The intensity I_2 after the first two polarizers

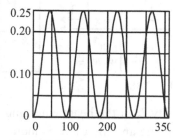

Fig. 3.37 The intensity I_3 after the three polarizers

$$E_2 = E_1 \cos\theta \qquad E_3 = E_2 \sin\theta = E_1 \cos\theta \sin\theta = \frac{1}{2}\sin(2\theta)$$

and

$$I_3 = \frac{1}{4}\sin^2(2\theta)$$

Maxima of $I_3 = 0{,}25$ are obtained for $\sin^2(2\theta) = 1$. Hence when

$$2\theta = m\frac{\pi}{2} \qquad m = 1,2,3,5 \qquad \theta = \frac{\pi}{4},3\frac{\pi}{4},5\frac{\pi}{4},7\frac{\pi}{4}$$

and minima when

$$2\theta = m\pi \qquad m = 0,1,2,3, \qquad \theta = 0,\frac{\pi}{2},\pi,3\frac{\pi}{2}$$

The maxima, the minima and all other values of I_3 in the range $(0, 2\pi)$ are immediately obtained with the matrix calculus (Fig. 3.37).

3.2.12 A half-wave plate

A beam of natural light, having intensity $I_1 = 1$ and moving along the z axis is incident on a first polarizer P_1, lying in the xy plane and whose transmission axis (or pass-plane) subtends angle $\theta = 45°$ with the x axis. The beam emerging from P_1 is incident on a half-wave plate $P_{h\text{-}w}$, parallel to P_1 and whose optic axis is parallel to the x axis. Emerging from the half-wave plate the beam is incident on a second polarizer P_2, parallel to P_1 and $P_{h\text{-}w}$, rotating on the z axis (Fig. 3.38). Give the plot of the inten-

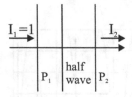

Fig. 3.38

sity I_2 of the beam emerging from P_2 when the angle θ between its pass-plane and the x axis is increasing from 0° to 360°. The polarizers and the half-wave plate, considered "ideal", don't absorb light. Use the matrix calculus.

Solution:

The beam emerging from the first polarizer doesn't change the intensity and reaches $P_{h\text{-}w}$ having the polarization form

$$v_1 = \frac{1}{\sqrt{2}} \begin{vmatrix} 1 \\ 1 \end{vmatrix}$$

The optic axis of the half-wave plate is parallel to x axis. The matrix of the half-wave plate is

$$m_1 = \begin{vmatrix} 1 & 0 \\ 0 & -1 \end{vmatrix}$$

The beam emerging from the half-wave plate has the linear polarization form

$$v_2 = m_1 v_1 = \frac{1}{\sqrt{2}} \begin{vmatrix} 1 \\ -1 \end{vmatrix} = \begin{vmatrix} \cos 45° \\ \sin(-45°) \end{vmatrix}$$

The electric field is rotated clockwise from the first to the fourth quadrant with reference to the xy plane.

The second polarizer, rotating on the z axis, subtends an angle θ variable from 0° to 360°. Thereby P_2 has the matrix

$$m_2 = \begin{vmatrix} C_1^2 & C_1 S_1 \\ C_1 S_1 & S_1^2 \end{vmatrix} \qquad C_1 = \cos\theta \qquad S_1 = \sin\theta$$

The beam emerging from P_2 has a variable form of polarization v_2 given by the product

$$v_2 = m_2 m_1 v_1$$

and a corresponding variable intensity I_2 following, as expected, the Malus'law (Fig. 3.39). There are two orientations (at 45° and 225°) for which the intensity is equal to zero and two (at 135° and 315°) where the intensity is equal to one.

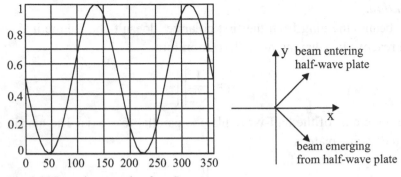

Fig. 3.39 Intensity emerging from P_2

3.2.13 A quarter-wave plate 1

A beam of natural light, having intensity $I_1 = 1$ and moving along the z axis is incident on a first polarizer P_1, lying in the xy plane and whose pass-plane is parallel to the x axis. The beam emerging from P_1 is incident on a quarter-wave plate P_{q-w}, parallel to P_1. The optic axis of P_{q-w} is parallel to the x axis. Emerging from the quarter-wave plate the beam is incident on a second polarizer P_2, parallel to P_1 and P_{q-w}. Give the plot of the intensity I_2 of the beam emerging from P_2 when the angle θ between its pass-plane and the x axis is increasing from $0°$ to $360°$. Next in time the direction of the pass plane of the first polarizer, P_1, passes from $0°$ to $45°$ (Fig. 3.40) leaving unchanged the direction of the optic axis of P_{q-w}. Give

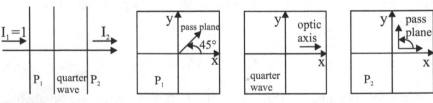

Fig. 3.40

the new plot of the intensity I_2 of the beam emerging from P_2 when the angle θ between its pass-plane and the x axis is increasing from $0°$ to $360°$. The polarizers and the quarter-wave plate, considered "ideal", don't absorb light. Use the matrix calculus.

Solution:

The beam emerging from the first polarizer doesn't change the intensity and reaches P_{q-w} having the polarization form

$$v_1 = \begin{vmatrix} 1 \\ 0 \end{vmatrix}$$

The optic axis of the quarter-wave plate is parallel to x axis. Its matrix is

$$m_1 = \begin{vmatrix} 1 & 0 \\ 0 & -i \end{vmatrix}$$

The polarization form emerging from the plate remains unchanged

$$v_e = \begin{vmatrix} 1 \\ 0 \end{vmatrix}$$

The second polarizer, rotating on the z axis and subtending an angle θ variable from 0° to 360°, will give rise to a beam that, according to the Malus' law, will have a maximum of intensity I_2 for 0° and 180° and a minimum fro 90° and 270° (Fig. 3.41).

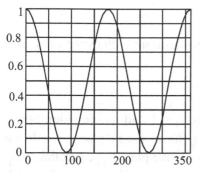

Fig. 3.41 Intensity I_2 when E and optic axis are parallel to the x axis

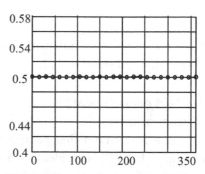

Fig. 3.42 Intensity I_2 when E is at 45° and optic axis is parallel to the x axis

If the direction of the pass plane of the first polarizer, P_1, passes from 0° to 45° the polarization form of the beam emerging from P_1 is

$$v_1 = \frac{1}{\sqrt{2}} \begin{vmatrix} 1 \\ 1 \end{vmatrix}$$

that reaches $P_{q\text{-}w}$ whose matrix is

$$m_1 = \begin{vmatrix} 1 & 0 \\ 0 & -i \end{vmatrix}$$

The beam emerging from the quarter-wave plate has the polarization form

$$v_s = m_1 v_1 = \frac{1}{\sqrt{2}} \begin{vmatrix} 1 \\ -i \end{vmatrix} = \begin{vmatrix} \cos\theta \\ \sin\theta e^{-i\frac{\pi}{2}} \end{vmatrix} \qquad \theta = 45°$$

The polarization form is circular ($\eta = 1$) and the handedness is left ($\varphi = \varphi_y - \varphi_x = -\pi/2$). Its intensity remains the same ($I_s = 1$).
The second polarizer, rotating on the z axis, subtends an angle θ variable from 0° to 360°. Thereby P_2 has the matrix

$$m_2 = \begin{vmatrix} C_1^2 & C_1 S_1 \\ C_1 S_1 & S_1^2 \end{vmatrix} \qquad C_1 = \cos\theta \qquad S_1 = \sin\theta$$

The beam emerging from P_2 will have the form of polarization v_3 given by the product

$$v_3 = m_2 m_1 v_1$$

but the corresponding intensity I_2 remains constant with every value of θ (Fig. 3.42) and equal to $I_1/2$.

3.2.14 A quarter-wave plate 2

A beam of natural light, having intensity $I_1 = 1$ and moving along the z axis is incident on a first polarizer P_1, lying in the xy plane. The beam emerging from P_1 is incident on a quarter-wave plate $P_{q\text{-}w}$, parallel to P_1. Emerging from $P_{q\text{-}w}$ the beam is incident on a second polarizer P_2, parallel to P_1 and $P_{q\text{-}w}$. The polarizer P_2 rotates around the z axis subtending an angle θ between its pass-plane and the x axis varying from 0° to 360°. Consider the positions of the electric field E emerging from P_1 and of the optic axis of $P_{q\text{-}w}$ described in Fig. 3.43. For this situation give the polarization form of the beam emerging from $P_{q\text{-}w}$ and plot the intensities I_2 of the beams emerging from P_2 when the angle θ between its pass-plane and the x axis is increasing from 0° to 360°. The polarizers and the quarter-wave plate, considered "ideal", don't absorb light. Use the matrix calculus.

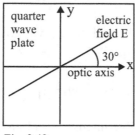

Fig. 3.43

Solution:
The beam emerging from P_1 has the polarization form

$$v_1 = \begin{vmatrix} \cos 30° \\ \sin 30° \end{vmatrix} = \begin{vmatrix} 0.866 \\ 0.500 \end{vmatrix}$$

The optic axis of the quarter-wave plate is parallel to x axis. Its matrix is

$$m_1 = \begin{vmatrix} 1 & 0 \\ 0 & -i \end{vmatrix}$$

The polarization form of the emerging from the plate is now elliptical and given by the product

$$v_e = m_1 v_1 = \begin{vmatrix} 0.866 \\ -0.500\,i \end{vmatrix} = \begin{vmatrix} 0.866 \\ 0.500 e^{-i\frac{\pi}{2}} \end{vmatrix}$$

The intensity of the beam emerging from quarter-wave plate remains equal to that of the beam entering the plate ($I_1 = 1$).
For the ellipticity, major and minor axis of the ellipse and the phase difference we have

$$\eta = \frac{0.500}{0.866} = 0.58 \qquad a = 0.866 \qquad b = 0.500 \qquad \varphi = -\frac{\pi}{2}$$

The handedness is left.
The second polarizer, rotating on the z axis, subtends an angle θ variable

with the x axis has the matrix

$$m_2 = \begin{vmatrix} C_1^2 & C_1 S_1 \\ C_1 S_1 & S_1^2 \end{vmatrix} \qquad C_1 = \cos\theta \qquad S_1 = \sin\theta$$

The beam emerging from P_2 will have a variable form of polarization v_3 given by the product

$$v_3 = m_2 m_1 v_1$$

with a corresponding variable intensity I_2 plotted in Fig. 3.44. The intensity I_2 is "perceived" by P_2 with a maximum (0.75) when its pass-plane subtends the angles $0°$ and $180°$, and a minimum (0.25) for $90°$ and $270°$.

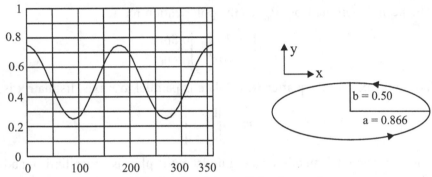

Fig. 3.44 I_2 when E is at 30°and optic axis is parallel to the x axis

3.2.15 A quarter-wave plate 3

A beam of natural light, having intensity $I_1 = 1$ and moving along the z axis is incident on a first polarizer P_1, lying in the xy plane. The beam emerging from P_1 is incident on a quarter-wave plate $P_{q\text{-}w}$, parallel to P_1. Emerging from $P_{q\text{-}w}$ the beam is incident on a second polarizer P_2, parallel to P_1 and $P_{q\text{-}w}$. The polarizer P_2 rotates on the z axis subtending an angle θ between its pass-plane and the x axis varying from 0 to 360°. Consider the positions of the electric field E emerging from P_1 and of the optic axis of $P_{q\text{-}w}$ described in Fig. 3.45. For this situation give the polarization form of the beam emerging from $P_{q\text{-}w}$ and plot the intensities I_2 of the beams emerging from P_2 when the angle θ between its pass-plane and

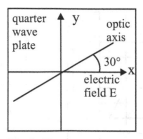

Fig. 3.45

the x axis is increasing from 0° to 360°. The polarizers and the quarter-wave plate, considered "ideal", don't absorb light. Use the matrix calculus.

Solution:
The beam emerging from P_1 has the polarization form

$$v_1 = \begin{vmatrix} 1 \\ 0 \end{vmatrix}$$

The optic axis of the quarter-wave plate is now at 30° from the x axis and the matrix of the plate becomes

$$m_1 = \begin{vmatrix} 0.75 - 0.25i & 0.433 + 0.433i \\ 0.433 + 0.433i & 0.25 - 0.75i \end{vmatrix}$$

The polarization form of the emerging from the plate is now

$$v_e = \begin{vmatrix} 0.75 - 0.25i \\ 0.433 + 0.433i \end{vmatrix}$$

The complex vector v_e is in rectangular form. Its polar form is

$$v_e = \begin{vmatrix} 0.79\,e^{-i0.32} \\ 0.61\,e^{i0.78} \end{vmatrix} \qquad \eta = \frac{0.61}{0.79} = 0.77 \qquad \varphi = 1.1\,\text{rad} = 63.4°$$

The polarization form of the beam is elliptical and handedness is right (Fig. 3.46). The intensity of the beam emerging from quarter-wave plate remains equal to that of the beam entering the plate ($I_1 = 1$).
The second polarizer, rotating on the z axis and subtending an angle θ variable with the x axis, has the matrix

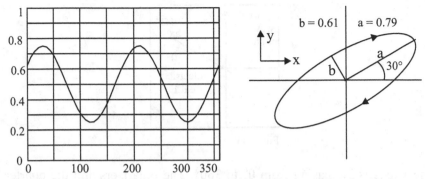

Fig. 3.46 Intensity I when the optic axis is at 30° and E is parallel to the x axis

$$m_2 = \begin{vmatrix} C_1^2 & C_1 S_1 \\ C_1 S_1 & S_1^2 \end{vmatrix} \qquad C_1 = \cos\theta \qquad S_1 = \sin\theta$$

The beam emerging from P_2 will have a variable form of polarization v_3 given by the product

$$v_3 = m_2 m_1 v_1$$

and a corresponding variable intensity I_2. In fact its intensity is "perceived" by P_2 as plotted in Fig. 3.46: a maximum (0.75) when its pass-plane subtends the angles 30° and 210°, and a minimum (0.25) for 120° and 300°.

Chapter 4

Interference

4.1 Main Laws and Formulae

4.1.1 Superposition of waves

In the preceding chapters we have had to do with a single (plane or sphe-rical) wave moving from a point A to a point B (Fig. 4.1) in many diffe-rent situations: there wasn't interference because there weren't two waves to be superposed. At most, in a retarder, there was a phase difference φ between the two components of the same wave but, clearly, no interfe-rence.

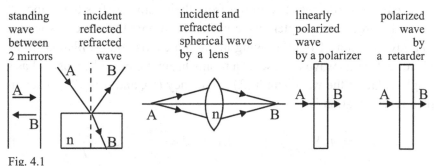

standing wave between 2 mirrors   incident reflected refracted wave   incident and refracted spherical wave by a lens   linearly polarized wave by a polarizer   polarized wave by a retarder

Fig. 4.1

In different situations (Fig. 4.2), where only reflection or both reflection and refraction are present, two or three rays reaching the point B corre-spond to a ray moving from the point A. In these conditions the interfe-rence occurs because the waves are superposed giving rise to a maximum, a minimum or an in-between intensity.

In the first case of Fig. 4.2 a ray moving from A reaches the semi-refle-

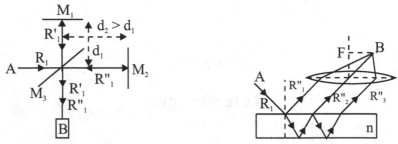

Fig. 4.2 Interference by reflection only (left) or by reflection and refraction (right)

cting mirror M_3 (also called a beam splitter). The two emerging rays are bounced back from M_1 and M_2 and, after traveling different distances ($d_2 > d_1$), reach again M_3 and then, superposed, the point B.

In the second case of Fig. 4.2 a ray A, incident on the boundary surface of a transparent substance, give rise, after refractions and reflections, to rays that converge, by means of a lens, to the point B, after traveling different distances.

A different case of interference, resulting from diffraction, will be discussed in the next chapter.

The superposition principle follows from the linear form of the Maxwell's differential equations (see Sec. 1.1.6): if two waves satisfy these equations, the wave resulting from their sum satisfies them, too.

As in the preceding chapters we consider harmonic the wave, associated to a ray or beam of light, moving along the z axis and oscillating in the plane xz. So this wave represents a monochromatic beam because it has a single value of the wavelength (λ), frequency (ν), and period T. Its complex representation is

$$E = Ae^{i(kz-\omega t)} \qquad A = E_M \qquad k = \frac{2\pi}{\lambda} \qquad \omega = \frac{2\pi}{T}$$

where A is the peak magnitude E_M. As a first simplest case consider two waves oscillating in the plane xz, having the same value for the peak (A), for the wavelength (λ), for frequency (ν), and period T and moving along the z axis but with a shift d between them (Fig. 4.3)

$$E_2 = Ae^{i(kz_2-\omega t)} \qquad E_1 = Ae^{i(kz_1-\omega t)} \qquad z_2 = z_1 + d$$

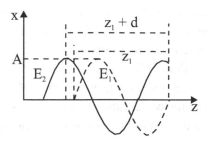

Fig. 4.3 Electric fields E_1 and E_2 in the plane xz at a fixed instant t

The shift d and the phase difference φ are related by the formula

$$\varphi = kd \qquad k = \frac{2\pi}{\lambda}$$

The wave resulting from the superposition the two waves E_2 and E_1

$$E = Ae^{i(kz_1-\omega t)} + Ae^{i[k(z_1+d)-\omega t]} = A(1+e^{ikd})e^{i(kz_1-\omega t)} = A(1+e^{i\varphi})e^{i(kz_1-\omega t)}$$

has the same value for the wavelength (λ), for frequency (ν), and period T but a different peak value (Fig. 4.4a) depending from $\varphi = kd$. The resulting intensity will be (Fig. 4.4b)

$$I = EE^* = 2A^2\left[1+\cos(\varphi)\right] = 4A^2\cos^2\left(\frac{\varphi}{2}\right)$$

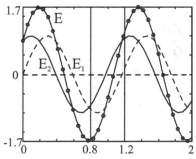

Fig. 4.4a The waves E_1, E_2 and $E=E_1+E_2$

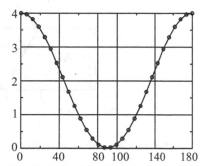

Fig. 4.4b The intensity of the wave E varying with the phase difference $\varphi/2$

Interference produces of a series of linear or circular fringes alternatively light and dark due to a maximum or a minimum of the intensity depending on whether the wave front of the initial beam is plane or spherical or from the geometry of the interference setup.

In problems dealing with interference the main task is to take count of the path difference d between the rays traveling from A to B or of the phase difference φ between the waves associated to the rays. The value of d depends on the geometry of the interference apparatus.

4.1.2 Interference by a thin homogeneous dielectric film

Fringes can be seen when a thin film of transparent material is viewed by reflected light. Consider a beam R, moving in the air, incident on the higher plane surface of a transparent film ($n>1$) of constant thickness t and the reflected rays R' and R" (Fig. 4.5). These rays R' and R" when reach the line ED have a path difference

$$d = n(AC + CD) - AE = 2nt\cos\alpha'$$

that, for low values of the angle of incidence and consequently lower values of the angle of refraction ($n>1$), becomes

$$d = 2nt$$

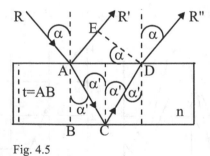

Fig. 4.5

The corresponding phase difference will be

$$\varphi = \pi + kd$$

where π is the phase change that occurs to the ray R' because the refle-

ction takes place in the less dense medium, the air (see Sec.1.2.21).
Real thin films, whose surface are not plane and whose thickness is not constant, are a layer of oil on water or on the surface of a road and a surface of a soap bubble. In these cases the fringes of interference are irregular, colored and difficult to calculate. Simple apparatus for interference due to reflected monochromatic light by a thin film of regular thickness are given in Fig. 4.6.

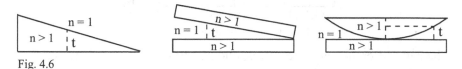

Fig. 4.6

4.2 Problems

4.2.1 A thin film on a lens

A lens of refractive index $n_2 = 1.85$, is covered on the superior surface with a very thin layer of a transparent substance, whose refractive index is n_1 and thickness is t. A beam of monochromatic light ($\lambda = 0.4\ \mu$) is incident normally on the layer and the lens (Fig. 4.7). Find the values n_1 and t for which the reflected ray is subject to a destructive interference

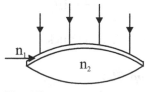

Fig. 4.7

Solution:
We assume that the same energy is conveyed by the rays reflected on the boundary surface between air and the layer and by the rays reflected on the boundary surface between the layer and the lens. So the two reflectivities are equal (Fig. 4.8)

$$R_1 = (\frac{n_1 - 1}{n_1 + 1})^2 = R_2 = (\frac{n_2 - n_1}{n_2 + n_1})^2 \qquad n_1 = \sqrt{n_2} = 1.36$$

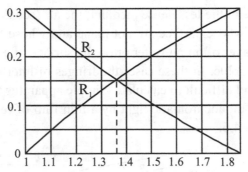

Fig. 4.8 Reflectivities functions of n_1

The path difference and the corresponding phase difference between the two reflected rays are

$$d = 2n_1 t \qquad \varphi = kd = \frac{2\pi}{\lambda} 2n_1 t$$

For a destructive interference the following condition must be verified

$$\varphi = \frac{2\pi}{\lambda} 2n_1 t = (2m+1)\pi \qquad t = \frac{(2m+1)\lambda}{4n_1}$$

For $m = 50$ we have

$$t = 7.4\,\mu$$

The layer is only a few wavelengths thick.

4.2.2 A wedge-shaped soap film

A monochromatic beam of light ($\lambda = 0.6\ \mu$) is incident on a wedge-shaped soap film ($n = 4/3$). Because the variation of the thickness t is very small the direction of the incident beam can be assumed normal to the surface of the bubble film (Fig. 4.9). Find the thickness difference between two consecutive light or dark fringes.

Solution:

The path difference and the phase difference are

$$d = 2nt \qquad \varphi = kd + \pi$$

where the additional phase shift π is due to the reflection at boundary surface (the first surface of the film) between a lower refractive index (air) and a higher refractive index (bubble).

The two reflected waves are

$$E_1 = Ae^{i\beta} \quad E_2 = Ae^{i(\beta+\varphi)}$$

and from their superposition we have

$$E = Ae^{i\beta}(1+e^{i\varphi}) = Ae^{i\beta}(1+e^{i(kd+\pi)}) = Ae^{i\beta}(1-e^{ikd})$$

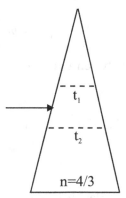

Fig. 4.9

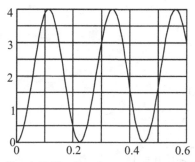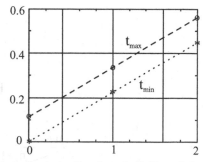

Fig. 4.10 The first three maxima and minima: relative intensity function of t (left); and the corresponding thicknesses t for m = 0, 1, 2 (right)

The corresponding absolute and relative intensities are (Fig. 4.10)

$$I = Ae^{i\beta}(1-e^{ikd})Ae^{-i\beta}(1-e^{-ikd}) = A^2 4\sin^2(\frac{kd}{2})$$

$$I_r = \frac{I}{A^2} = 4\sin^2(\frac{kd}{2})$$

For the following conditions of the path shift d and of the thickness t are obtained the maxima

$$d_{max} = (2m+1)\frac{\lambda}{2} \qquad t_{max} = (2m+1)\frac{\lambda}{4n}$$

and the minima

$$d_{min} = m\lambda \qquad t_{min} = m\frac{\lambda}{2n}$$

The thickness differences between two consecutive light or dark fringes are

$$t_{max2} - t_{max1} = \frac{\lambda}{2n} = 0.225\,\mu \qquad t_{min2} - t_{min1} = \frac{\lambda}{2n} = 0.225\,\mu$$

4.2.3 Newton's rings 1

A planoconvex lens lies on a flat surface of a plate. Both have the same refractive index n_1. The radius of the lens is $R = 6.2$ m. A beam of monochromatic light ($\lambda = 0.643\ \mu$), incident on the plane surface of the lens,

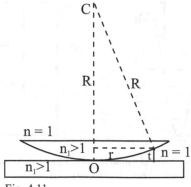

Fig. 4.11

forms a system of circular fringes around the point of contact O (Fig. 4.11). Are the fringes in some way dependent from the refractive index n_1? Give a simplified ray tracing outlining where the two reflected rays, that causes interference, start. For the first ten dark circular fringes find the corresponding thicknesses t, the radii r and the differences between a value of r and the preceding one. Give plots of the radii r, of their differences and of the thicknesses t as functions of the order m of fringes.

Solution:
Clearly the fringes are fully independent from the refractive indices of the lens and the flat surface. They are dependent only from R. The simplified ray tracing is given in Fig. 4.12. The condition for dark fringes is

$$k\,2t + \pi = (2m+1)\pi \qquad \frac{2\pi}{\lambda}2t = 2m\pi \qquad t = m\frac{\lambda}{2}$$

where π on the left side is added because the ray B" (Fig. 4.12) is reflected on the boundary surface air-plate.

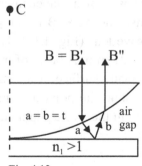

Fig. 4.12

From geometry (Fig. 4.11) t is also given by the following formula

$$t = \frac{1}{2}\frac{r^2}{R}$$

if we assume $r/R < 1$.
From the two previous formulae we have

$$\frac{1}{2}\frac{r^2}{R} = m\frac{\lambda}{2} \qquad r = \sqrt{m\lambda R}$$

The requested values are in Table 4.1 and depicted in Fig. 4.13.

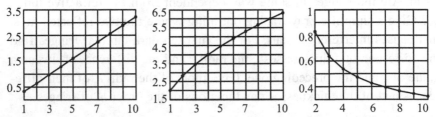

Fig. 4. 13 Rings thickness t (left, in μ), rings radii r (center, ,in mm), rings radii difference (right, in mm). The orders m of the dark fringes in abscissae

Table 4.1

m	1	2	3	4	5	6	7	8	9	10
t (μ)	0.32	0.64	0.96	1.29	1.61	1.93	2.25	2.57	2.89	3.22
r (mm)	2.0	2.8	3.5	4.0	4.5	4.9	5.3	5.7	6.0	6.3
diff (mm)		0.83	0.64	0.54	0.47	0.43	0.39	0.37	0.34	0.32

4.2.4. The Newton's rings 2

The equipment A is composed of a planoconvex lens lying on another reversed planoconvex lens. The other B is composed of a planoconvex lens lying on a planoconcave lens (Fig. 4.14). The radii of the lenses are $R_1 = 1555$ mm and $R_2 = 6221$ mm. A beam of monochromatic light ($\lambda = 0.643$ μ), incident on the plane surface of the superior lens, forms a system of circular fringes around the point of contact O. How change the radii r and the thicknesses t in the two equipments varying the order m of the fringes?

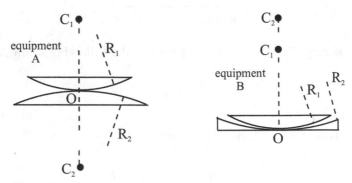

Fig. 4. 14

Solution:

In both equipments, by their geometry (see Problem 4.2.3), the thicknesses of the two lenses are

$$t_1 = \frac{1}{2}\frac{r_1^2}{R_1} \qquad t_2 = \frac{1}{2}\frac{r_2^2}{R_2}$$

In the equipment A the overall thickness is (Fig. 4.15A)

$$t_A = t_1 + t_2 = \frac{r_A^2}{2}\left(\frac{1}{R_1} + \frac{1}{R_2}\right)$$

The condition for a dark fringe is in both equipments

$$t = m\frac{\lambda}{2}$$

hence for the equipment A

$$r_A = \sqrt{m\lambda\frac{R_1 R_2}{R_1 + R_2}}$$

In the equipment B the overall thickness is (Fig. 4.15B)

$$t_B = t_1 - t_2 = \frac{r_B^2}{2}\left(\frac{1}{R_1} - \frac{1}{R_2}\right)$$

and the corresponding radii of the fringes are

$$r_B = \sqrt{m\lambda\frac{R_1 R_2}{R_1 - R_2}}$$

Therefore for a fixed order m the value of t is the same (it's a function only of λ) but for the radii of a fringe of order m, in the two apparatus, is always $r_B > r_A$ (Fig. 4.15 and Fig. 4.16).

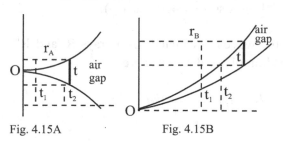

Fig. 4.15A   Fig. 4.15B

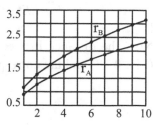

Fig. 4.16

4.2.5 Michelson 1

A monochromatic beam ($\lambda = 0.643 \ \mu$) is incident on the beam splitter M_3 of the Michelson apparatus (Fig. 4.17). The ray R is incident on the sur-

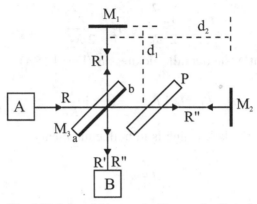

Fig. 4.17 Only axial beams incident and reflected under 45° with normal to M_3 are shown. The beams with slightly larger angles than 45° are leading to interference with circular fringes

face a-b of M_3, where it is split into the reflected ray R' and the transmitted ray R". The two mirrors, M_1 and M_2, return the rays, R' and R", superposed, to B. From the initial condition, $d_1 = d_2$, with the mirror M_1 fixed and distance d_1 constant, the mirror M_2 is moved increasing the distance d_2 until 1000 dark circular fringes are counted on B. Find the displacement $d = d_2 - d_1$ when this counting is over.

Solution:

In Sec. 4.1.1 was found the intensity of two superposed waves with a phase difference $\varphi = kd$ where d was the distance between the two waves

$$I = EE^* = 2A^2\left[1 + \cos(\varphi)\right] = 4A^2 \cos^2(\frac{\varphi}{2})$$

Now the distance between the waves associated to the rays R' and R" (Fig. 4.18) and the corresponding phase difference are

$$2d = 2(d_2 - d_1) \qquad \varphi = 2kd$$

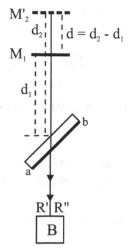

Fig.. 4.18 To observer at B rays
R' and R" appear to come from
M_1 and virtual mirror M'_2

Maxima for I are obtained when

$$\frac{\varphi_{max}}{2} = \frac{2kd_{max}}{2} = \frac{2\pi}{\lambda}d_{max} = m\pi \qquad d_{max} = m\frac{\lambda}{2} \qquad m = 0,1,2,..$$

and minima when

$$\frac{\varphi_{min}}{2} = \frac{2kd_{min}}{2} = (2m+1)\frac{\pi}{2} \qquad d_{min} = (2m+1)\frac{\lambda}{4} \qquad m = 0,1,2,...$$

When 1000 dark fringes have been counted the displacement d is 321.7 μ = 0.3 mm.

4.2.6 Michelson 2

A beam, composed of the two sodium D-lines ($\lambda_1 = 0.58900$ μ and $\lambda_2 = 0.58958$ μ), is incident on the splitter M_3 of the Michelson apparatus (Fig. 4.17). We assume both D-lines can be considered monochromatic or harmonic waves. Both are split (in the Michelson apparatus are moving four waves) and then the two waves of wavelength λ_1 are reaching B superposed and so happen to the two waves of wavelength λ_2. Find the whole intensity observed in B and plot how it is varying with $d = d_2 - d_1$ (the displacement between the two mirrors M_1 and M_2). Find the number m of fringes that must be counted to obtain the first minimum of the intensity and the visibility given by the Michelson formula

$$V = \frac{|I_{r\,max} - I_{r\,min}|}{I_{r\,max} + I_{r\,min}}$$

Explain why the intensity isn't constant and varies with d.

Solution:

In the preceding problem we found for the intensity for a beam, of single wavelength λ, incident on the Michelson apparatus

$$I = 2A^2\left[1 + \cos(\varphi)\right] \qquad I_r = \frac{I}{A^2} = 2\left[1 + \cos(\varphi)\right] \qquad \varphi = 2kd$$

For the two wavelengths λ_1 and λ_2 the corresponding relative intensities will be

$$I_{r1} = 2(1 + \cos\varphi_1) \qquad I_{r2} = 2(1 + \cos\varphi_2)$$

where (Fig. 4.19)

$$\varphi_1 = 2k_1 d \qquad \varphi_2 = 2k_2 d$$

with

$$k_1 = k + \frac{\Delta k}{2} \qquad k_2 = k - \frac{\Delta k}{2}$$

The value of k_1 is obtained adding and that of k_2 by subtracting the following formulae

$$k = \frac{k_1 + k_2}{2} \qquad \Delta k = k_1 - k_2 \qquad \frac{\Delta k}{2} = \frac{k_1 - k_2}{2}$$

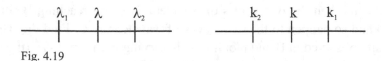

Fig. 4.19

The whole intensity observed in B of the Michelson apparatus will be (Fig. 4.20)

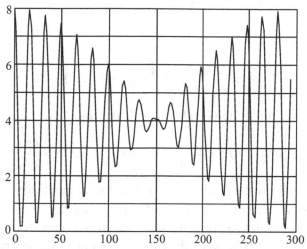

Fig. 4.20 The whole intensity function of d (μ)

$$I_r = I_{r1} + I_{r2} = 2\left[1 + \cos(k + \frac{\Delta k}{2})2d)\right] + 2\left[1 + \cos(k - \frac{\Delta k}{2})2d)\right]$$

that, using trigonometric formula for a cosine of the sum of two angles, becomes

$$I_r = 4[1 + A\cos(2kd)] \qquad A = \cos(d\Delta k)$$

The derivatives of the first and second order of I_r are

$$\frac{dI_r}{dk} = -8Aksen(2kd) \qquad \frac{dI_r}{dk} = 0 \;\rightarrow\; 2kd = m\pi$$

$$\frac{d^2 I_r}{dk^2} = -16Ak^2\cos(2kd) = \begin{cases} <0 & 2kd = 2m\pi & d_{max} = m\frac{\lambda}{2} \\ >0 & 2kd = (2m+1)\pi & d_{min} = (2m+1)\frac{\lambda}{4} \end{cases}$$

Obviously we find again the standard condition for the maxima and the minima of the intensity of the composite beam. Hence

$$I_{r\,max} = 4[1 + A\cos(2m\pi)] \qquad I_{r\,min} = 4[1 - A\cos((2m+1)\pi)]$$

or for $m = 0, 1, 2, 3, \ldots$

$$I_{r\max} = 4(1 + A) \qquad I_{r\min} = 4(1 - A)$$

The visibility is

$$V = \frac{\left|I_{r\max} - I_{r\min}\right|}{I_{r\max} + I_{r\min}} = \left|\cos(d * \Delta k)\right|$$

The first minimum of the visibility, $V = 0$, requires

$$d * \Delta k = \frac{\pi}{2} \qquad d* = \frac{\pi}{2\Delta k} = 149.68\,\mu \qquad m* = \frac{2d*}{\lambda} = 508$$

The distance $d*$ for the minimum of V is equal to the distance necessary for a shift of $\lambda/2$ between the peak of superposed ray of wavelength λ_1 and that of wavelength λ_2 (Fig. 4.21). In fact there is necessary a number of small $\lambda_2 - \lambda_1$ shifts between the two rays to obtain the entire displacement $\lambda/2$. This number is

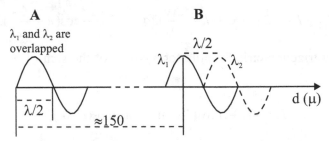

Fig. 4.21 In position **A** the intensities of the two rays are added. In **B** they aren't and the whole intensity is half of that in **A**

$$\frac{\lambda/2}{\lambda_2 - \lambda_1} = 508$$

that corresponds to the passage of $m*$ bright fringes in B. Hence

$$d* = 508\frac{\lambda}{2} = 149.68\,\mu$$

4.2.7 Multiple-beam interference

A beam of monochromatic light ($\lambda = 0.435$ μ) is incident with an angle θ on a plane, thin, parallel plate of thickness $t = 0.5$ mm and refractive index $n = 1.4$ (Fig. 4. 22). Disregarding initial reflection on A and suc-

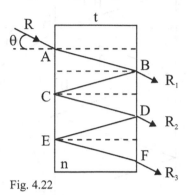

Fig. 4.22

ceeding back refractions on C and E, parallel rays (R_1, R_2, R_3, . . .) emerge from the points B, D, F, . . . due to the repeated forth and back reflections on the internal surface of the plate. We assume there isn't energy absorption by the plate. Find the highest allowable order m of interference and the corresponding value of the angle of refraction θ''. Plot θ'' as a function of m. Find the values of the angle of refraction θ'' and of the corresponding m for $\theta = 14.1°$.

Solution:

The distance (Fig. 4. 23) between R_2 and R_1 and between R_3 and R_1 are

$$d = 2n\,BC - BG \qquad d' = 4n\,BC - 2BG = 2(2n\,BC - BG) = 2d$$

Using the simple relations

$$BC = \frac{t}{\cos\theta'} \qquad BB' = BC\sin\theta' \qquad BG = 2\,BB'\sin\theta \qquad \sin\theta = n\sin\theta'$$

the optical path differences become

$$d = 2nt\cos\theta' \qquad d' = 2(2nt\cos\theta') = 2d$$

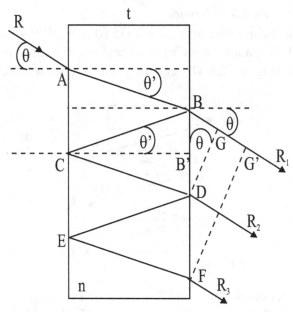

Fig. 4. 23

Hence the corresponding phase differences are

$$\varphi = kd \qquad \varphi' = kd' = 2\varphi$$

Calling now ρ and τ the reflectivity and refractivity for natural light (see Sec. 3.1.3 where they were called R* and T*) the values

$$\varepsilon_r = \sqrt{\rho} \qquad \varepsilon_t = \sqrt{\tau}$$

are the factors by which amplitude of a wave associated to an incident ray must be multiplied to obtain the corresponding amplitudes of the waves associated to the reflected and refracted rays.

Hence we have (Fig. 4.24) for the amplitudes of the waves associated to the transmitted rays R_1, R_2, R_3

$$a_1 = \varepsilon_t^2 a = \tau a \qquad a_2 = \varepsilon_r^2 \varepsilon_t^2 a = \rho \tau a \qquad a_3 = \varepsilon_r^4 \varepsilon_t^2 a = \rho^2 \tau a$$

$$a_p = \varepsilon_r^{2(p-1)} \varepsilon_t^2 a = \rho^{(p-1)} \tau a$$

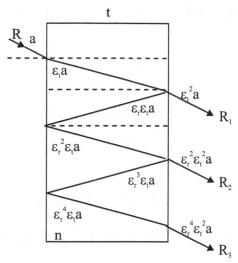

Fig. 4.24

Therefore if the wave, assumed harmonic, for the incident ray is

$$y = ae^{i\alpha}$$

for the rays R_1, R_2, R_3, ... R_p we can write

$$y_1 = \tau ae^{i\alpha} \qquad y_2 = \rho\tau ae^{i(\alpha+\varphi)} \qquad y_3 = \rho^2 \tau ae^{i(\alpha+2\varphi)}$$

$$y_p = \rho^{(p-1)} \tau ae^{i(\alpha+(p-1)\varphi)}$$

Then the resulting wave will be

$$y_r = \tau ae^{i\alpha} + \rho\tau ae^{i(\alpha+\varphi)} + \rho^2 \tau ae^{i(\alpha+2\varphi)} +$$

$$\dots + \rho^{(p-1)} \tau ae^{i(\alpha+(p-1)\varphi)} =$$

$$= \tau ae^{i\alpha} (1 + \rho e^{i\varphi} + \rho^2 e^{i2\varphi} + \dots + \rho^{(p-1)} e^{i(p-1)\varphi})$$

and the corresponding relative intensity (see Appendix 2)

$$I_r = \frac{\tau^2}{1+\rho^2 - 2\rho \cos\varphi}$$

that has the maximum

$$I_{r\max} = \frac{\tau^2}{(1-\rho)^2} = 1$$

This requires (Fig. 4.25)

$$\varphi = 2m\pi = k2nt\cos\theta' \qquad \theta' = \arccos(\frac{m\pi}{knt})$$

The least value of the angle of refraction (θ' =0°) requires cos(θ') = 1. Hence

$$\frac{m\pi}{knt} = 1 \qquad m = \frac{knt}{\pi} = 3218$$

For $\theta = 14.1°$ we have $\theta' = 10.1°$ and $m = 3169$.

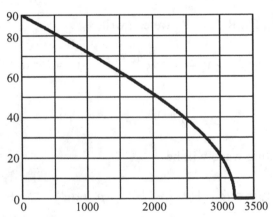

Fig. 4.25 The angle θ' function of m

4.2.8 Fabry-Perot interferometer 1

A monochromatic light ($\lambda = 0.435$ µ) source emitting a beam into a solid angle is incident on a Fabry-Perot interferometer, composed of two plane, parallel glass plates S and S' at a distance $t = 1$ mm apart (Fig. 4. 26 and

Fig. 4.27). The interior surfaces of the plates, coated with a silvered film, are partially reflecting. Disregarding initial reflection on A and succeeding back transmissions on C and E, parallel rays (R_1, R_2, R_3, . . .) emerge

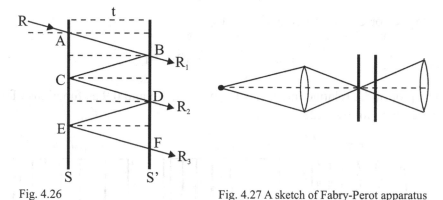

Fig. 4.26   Fig. 4.27 A sketch of Fabry-Perot apparatus

from the points B, D, F, . . . due to the repeated forth and back reflections on the internal surface of the plates. We assume there isn't energy absorption by the plate. We call now ρ the reflectivity for natural light (see Sec. 3.1.3 where it was called R*). For values of the reflectivity $\rho = 0.3$, 0.6 and 0.9 find and plot the relative intensity as a function of the angle of incidence θ varying in the interval (0.5, 3)°. Verify how minima and maxima of the intensity vary changing the value of the reflectivity.

Solution:
The steps to find the final relative intensity are equal to that used in Sec. 4.2.7.

$$I_r = \frac{(1-\rho)^2}{1+\rho^2 - 2\rho\cos\varphi}$$

but with the phase shift now defined as

$$\varphi = k2d\cos\theta$$

The maxima and the minima are now

$$I_{r\,\text{max}} = \frac{\tau^2}{1+\rho^2 - 2\rho} = \frac{\tau^2}{(1-\rho)^2} = 1$$

and

$$I_{r\,min} = \frac{\tau^2}{1+\rho^2+2\rho} = \frac{\tau^2}{(1+\rho)^2} = \frac{(1-\rho)^2}{(1+\rho)^2} < 1$$

with the corresponding conditions for φ

$$\varphi_{max} = m\,2\pi \qquad \varphi_{min} = (2m+1)\pi$$

The plot of the relative intensities for $\rho = 0.3$, 0.6 and 0.9 as functions of θ are given in Fig. 4.28.

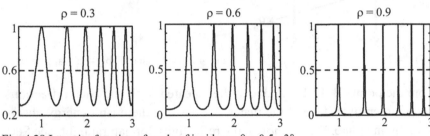

Fig. 4.28 Intensity function of angle of incidence $\theta = 0.5 - 3°$

The values of the minima for the three values of ρ are respectively 0.3, 0.1 and 0.0.

4.2.9 Fabry-Perot interferometer 2
A beam, composed of the two sodium D-lines ($\lambda_1 = 0.58900$ μ and $\lambda_2 = 0.58958$ μ), is incident on the Fabry-Perot apparatus. We assume both D-lines can be considered as harmonic waves.
The two plates S and S' are at a distance $t = 2$ mm apart. We assume there isn't energy absorption by the plate and the reflectivity is $\rho = 0.85$. Change the formula for I_r (see previous problem) expressing cosine in the denominator as a sine function. Plot their values as a function of the angle of incidence θ varying in the interval (0.2, 1.0)° for both values of wavelengths. Calculate the angular positions of the two maxima and of the minima to left and to right of the maxima and the corresponding orders m of the fringes.

Solution:

Using the trigonometric relation

$$\cos\varphi = 1 - 2\sin^2\frac{\varphi}{2}$$

the formula for relative intensity becomes

$$I_r = \frac{(1-\rho)^2}{1+\rho^2-2\rho+4\rho\sin^2\frac{\varphi}{2})} = \frac{(1-\rho)^2}{(1-\rho^2)+4\rho\sin^2\frac{\varphi}{2})} =$$

$$= \frac{1}{1+\dfrac{4\rho}{(1-\rho)^2}\sin^2(\frac{\varphi}{2})} = \frac{1}{1+G\sin^2(\frac{\varphi}{2})} \qquad G = \frac{4\rho}{(1-\rho)^2}$$

The plot of this function for λ_1 and λ_2 is given in Fig. 4.29. The angular positions of the minima to the left of the maxima obviously are $\theta = 0$. Their orders $m_1 = 6791$ and $m_2 = 6784$ respectively for λ_1 and λ_2 are found using the formulae ($\cos\theta = 1$ for $\theta = 0$)

$$k2d\cos\theta = (2m+1)\pi \qquad \rightarrow \qquad m_1 = \frac{2k_1d - 1}{2} \qquad m_2 = \frac{2k_2d - 1}{2}$$

The angular positions of the maxima for λ_1 and λ_2 are

$$\theta_1 = 0.41° \qquad\qquad \theta_2 = 0.69°$$

and their orders remain $m_1 = 6791$ and $m_2 = 6784$.
The angular positions of the minima to the right of the maxima are

$$\theta_1 = 0.81° \qquad\qquad \theta_2 = 0.98°$$

and their orders are now $m_1 = 6790$ and $m_2 = 6783$ respectively for λ_1 and λ_2.

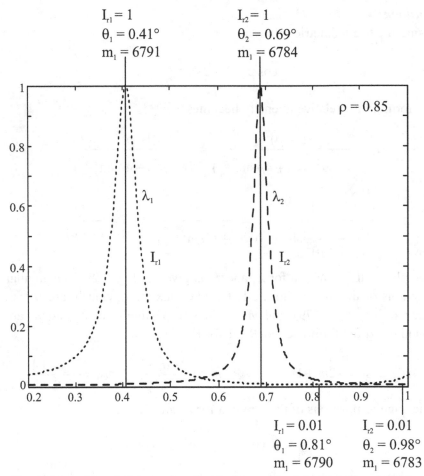

Fig. 4.29 Intensities functions of angle of incidence $(\theta = 0.2 - 1°)$ for $\lambda_1 = 0.58900\ \mu$ and $\lambda_2 = 58958\ \mu$

4.2.10 Fabry-Perot interferometer 3

A beam, composed of the two sodium D-lines ($\lambda_1 = 0.58900\ \mu$ and $\lambda_2 = 0.58958\ \mu$), is incident on the Fabry-Perot apparatus. We assume both D-lines can be considered as harmonic waves. The two plates S and S' are at a distance $t = 2$ mm apart. We assume there isn't energy absorption by the plate and the reflectivity is $\rho = 0.95$. Plot the intensities as a function of the angle of incidence θ varying in the interval $(0.2, 1.0)°$ for both values of wavelengths. Calculate the angular positions of the two maxima and of the two minima to right of the maxima. Compare these results with those of the previous problem and verify if the Rayleigh's criterion is satisfied. Find the expected value of t that could fulfill this criterion.

Solution:
The results for $\rho = 0.95$ are given in Fig. 4.30 (for the maxima) and in Fig. 4.31 (for the minima). The Rayleigh's criterion isn't fulfilled neither with $\rho = 0.85$ nor with $\rho = 0.95$ even if deviation is very small ($\Delta\theta = 0.3°$).

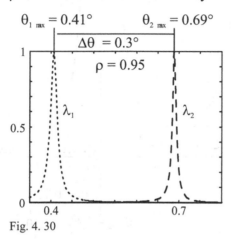

Fig. 4. 30

The expected value of t fulfilling the Rayleigh's criterion can be found with the following arguments. The conditions for the maximum of the order m_1 and the next minimum of order $m_1 - 1$ for the wavelength λ_1 are

$$2d\cos\theta_1 = m_1\lambda_1$$

$$2d\cos\theta_2 = [2(m_1 - 1) + 1]\frac{\lambda_1}{2}$$

$$2d\cos(\theta_1 + \Delta\theta_{Mm}) = (2m_1 - \frac{1}{2})\lambda_1$$

The last formula can also be written, remembering that θ_1 and $\Delta\theta_{Mm}$ are small

$$2d[\cos\theta_1 - \theta_1\Delta\theta_{Mm}] = (m_1 - \frac{1}{2})\lambda_1$$

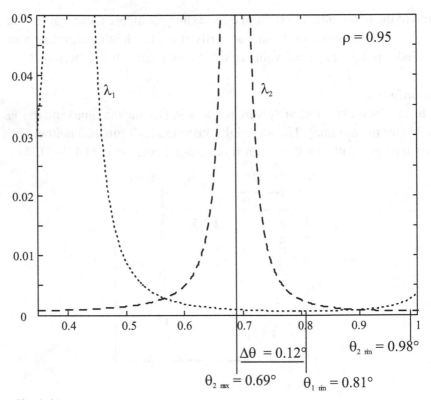

Fig. 4. 31

Subtracting the formulae

$$2d\cos\theta_1 - 2d\theta_1\Delta\theta_{Mm} - 2d\cos\theta_1 = (m_1 - \frac{1}{2})\lambda_1 - m_1\lambda_1$$

we obtain

$$\Delta\theta_{Mm} = \frac{\lambda_1}{4d\theta_1}$$

The condition for the maximum of the adjacent bright fringe, of the order m_2 and wavelength λ_2, is

$$2d\cos\theta_2 = m_2\lambda_2 \qquad 2d\cos(\theta_1 + \Delta\theta_{MM}) = m_2\lambda_2$$

or

$$2d\cos\theta_1 + 2d\theta_1\Delta\theta_{MM} = m_2\lambda_2$$

The highest value of m_2 requires $\cos\theta_2 = 1$ for the central dark fringe (but this m_2 is the order of the subsequent bright fringe, too)

$$k_2 2d = (2m_2 + 1)\pi \qquad m_2 = \frac{t}{\lambda_2} - 0.5$$

Hence

$$2t\cos\theta_1 + 2t\theta_1\Delta\theta_{MM} = m_2\lambda_2 = (\frac{t}{\lambda_2} - 0.5)\lambda_2 = t - \frac{\lambda_2}{2}$$

$$\Delta\theta_{MM} = \frac{1}{2t\theta_1}(t - \frac{\lambda_2}{2} - 2t\cos\theta_1) = \frac{1}{\theta_1}(\frac{1}{2} - \frac{\lambda_2}{4t} - \cos\theta_1)$$

Equating the values $\Delta\theta_{Mm}$ and $\Delta\theta_{MM}$, according to the Rayleigh's criterion we would have

$$\frac{1}{\theta_1}(\frac{1}{2} - \frac{\lambda_2}{4t} - \cos\theta_1) = \frac{\lambda_1}{4t\theta_1} \qquad 2t - \lambda_2 + 4t\cos\theta_1 = \lambda_1$$

From the last equation we obtain, assuming (see θ_1 in Figs 4.30 and 4.31) $\cos\theta_1 = 1$, a value for the thickness

$$t = \frac{\lambda_1 + \lambda_2}{6} = 0.2\,\mu$$

that isn't usable with the Fabry-Perot apparatus.

4.2.11 Fabry-Perot interferometer 4

A beam of a monochromatic light ($\lambda_1 = 0.58900\ \mu$), is incident on the Fabry-Perot apparatus. The two plates S and S' are at a distance $t = 2$ mm apart. We assume there isn't energy absorption by the plate and the reflectivity is $\rho = 0.60$. Plot the relative intensity as a function of the angle of incidence θ varying in the interval $(0, 1.3)°$. Calculate the angular positions, to the right of the first two maxima, where the relative intensity is equal to 1/2.

Solution:

If $\Delta\varphi$ is the phase difference between the points, where $I_r = 1/2$, on either side of a maximum, where $I_r = 1$, it follows

$$\varphi = 2m\pi \pm \frac{\Delta\varphi}{2}$$

We use this value of φ in the formula that gives I_r. Putting $I_r = 0.5$ and remembering trigonometric formula for sine of a sum of two angles and that $\Delta\varphi$ is small we have

$$\frac{1}{2} = \frac{1}{1 + G\sin^2(\varphi)} = \frac{1}{1 + G\sin^2(\frac{\Delta\varphi}{2})} = \frac{1}{1 + G(\frac{\Delta\varphi}{2})^2} \qquad G = \frac{4\rho}{(1-\rho)^2}$$

Hence

$$\Delta\varphi = \frac{2}{\sqrt{G}} = \frac{(1-\rho)^2}{\sqrt{\rho}}$$

The corresponding angle θ_{mr}, on the right of the maximum, where $I_r = 1/2$, will be

$$\varphi_{mr} = 2m\pi + \frac{\Delta\varphi}{2} = k2d\cos(\theta_{mr}) \qquad \theta_{mr} = \arccos(\frac{2m\pi + \frac{\Delta\varphi}{2}}{k2d})$$

For the first maxima the orders are, respectively, $m_1 = 6791$ and $m_2 = 6790$. We have

$$G = 15 \qquad \Delta\varphi = 1.03$$

and for the requested angular positions (Fig. 4. 32)

$$\theta_{m_1 r} = 0.5° \qquad \theta_{m_2 r} = 1.1°$$

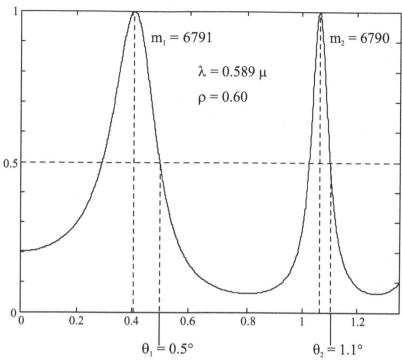

Fig. 4. 32

Chapter 5

Diffraction

5.1 Main Laws and Formulae

5.1.1 Introduction

When a beam of light crosses the edge or the opening of an opaque surface the propagation is not truly linear. The shadow on a screen is not a geometrical and definite image of the edge or the opening. Light penetrates into the region of the shadow that ought to be completely dark if the light propagation was really linear. Besides, dark and bright fringes appear near the blurred black image of the edge or the opening. These deviations from the geometrical model of the linear propagation of a light ray are examples of diffraction. Many renowned physicists (Huygens, Fresnel, Fraunhofer, Kirchhoff, Rayleigh, Sommerfeld, Airy, etc.) have attempted to explain diffraction on a wave-theoretical basis obtaining rigorous solutions only for few diffraction subjects. For the presence of very difficult mathematical problems, approximate methods are used in most cases of practical interest.

The presence of fringes, as a result of the diffraction process, implies interference too. Therefore most of the mathematical tools, used in the previous chapter, are needed also now. As in the previous chapter we don't discuss the matter of coherence. The problems we are dealing with suppose that the light is strictly monochromatic. Hence the light is assumed coherent and the phase difference constant between the waves moving from the source, where they originate, to the final place, where they superpose and the interference fringes appear.

A preeminent aspect of diffraction is the presence of its effects in the process of image formation by all optical instruments, apparatus or vision devices (including eyes, too). Diffraction sets an inescapable limit to the sharpness, in the focus space, of the optical image or, as a special case, to the accuracy of measurements and observation of long-distant objects. An old criterion was proposed by Rayleigh to define the resolving power of the fringes. But different criteria can be used.

A usual distinction is made between Fresnel and Fraunhofer diffraction. The former case occurs when the source and/or the final screen are close to the diffracting device: the result is a blurred image of the initial object. The latter case occurs when both incident and diffracted waves can be considered plane because the distance from the source to the diffracting device and the distance from this device to the final screen are large enough: the diffraction result is a series of fringes whose intensity has values varying, according to some laws, from a maximum to a minimum. When the aperture, as a special case, is a regular and linear slit or is composed by two (the Young's device) or by a large number N (a device called "grating") of parallel slits the law that regulates the intensity (see next Section) is the square of the sinc (α) function (sin(α)/α) (Fig. 5.1).

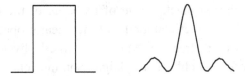

Fig. 5.1 The "top-hat" function (left) represents the intensity entering a single slit. The pattern of its Fraunhofer diffraction is represented by the square of a "sinc" function (right)

5.1.2 Fraunhofer diffraction from one or N linear slits

In Fig. 5.2 is given a simplified apparatus for diffraction using two slits: h is their width, d is the distance between them and θ is an angular position giving rise to the diffraction on a single point of the screen A'-B'. In Fig. 5.2 are shown, emerging from the two slits, only two of a large number of rays whose waves superpose in one point, at angular position θ on the screen. Varying θ other points of the diffraction image are created. A

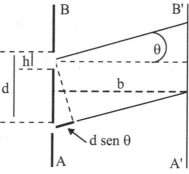

Fig. 5.2

set of fringes of variable intensity appears, as a final result, on the screen A'-B'. The formula of relative intensity for N slits is

$$I_r = (\frac{\sin\alpha}{\alpha})^2 \ (\frac{\sin N\beta}{\sin\beta})^2$$

where

$$\alpha = \pi\frac{h}{\lambda}\sin\theta \qquad \beta = \pi\frac{d}{\lambda}\sin\theta$$

(see Sec. 5.2.1 for the first factor of I_r and Appendix 2 for the second). The variables α and β, for assigned values of h, d and λ are function of θ. Normalizing the formula (and so the maximum of I_r is always one) and emphasizing the relation between I_r and θ we rewrite the previous formula as

$$I_r(\theta) = (\frac{\sin\alpha}{\alpha})^2 \ (\frac{\sin N\beta}{N\sin\beta})^2$$

For $N = 1$ we have

$$I_r = (\frac{\sin\alpha}{\alpha})^2$$

and for $N = 2$

$$I_r = (\frac{\sin\alpha}{\alpha})^2 \ (\frac{\sin 2\beta}{2\sin\beta})^2 = (\frac{\sin\alpha}{\alpha})^2 \ (\frac{2\sin\beta\cos\beta}{2\sin\beta})^2 = (\frac{\sin\alpha}{\alpha})^2 \cos^2\beta$$

For N = 1 we have the following conditions:
the *central maximum* when $\alpha = 0$ and then $\theta = 0$; the *other maxima* when

$$\alpha = \frac{\pi}{\lambda}\, h\sin\theta = (2m+1)\frac{\pi}{2} \qquad h\sin\theta = (2m+1)\frac{\lambda}{2} \qquad m=0,1,...$$

the *minima* when

$$\alpha = \frac{\pi}{\lambda}\, h\sin\theta = m\pi \qquad h\sin\theta = m\lambda \qquad m=1,2,...$$

For N > 1 we have the following *condition for the maxima*

$$\beta = \frac{\pi}{\lambda}\, d\sin\theta = m\pi \qquad d\sin\theta = m\lambda \qquad m=0,1,.2,...$$

The *minima* for **two** slits occur when

$$\beta = \frac{\pi}{\lambda}\, d\sin\theta = (2m+1)\frac{\pi}{2} \qquad d\sin\theta = (2m+1)\frac{\lambda}{2} \qquad m=0,1,.2,...$$

For N slits between two adjacent maxima there are N-2 secondary maxima, whose intensity is very small, and N-1 minima. Their formulae are unimportant for our problems.

The angular positions of the bright fringes are the same for 2, 3 ... N slits. But the relative intensity of the bright fringes is equal to N^2 and their thickness is inversely proportional to N. For very large number of N (> 1000) the bright fringes are therefore called "lines".

For N slits the *resolving power* of two lines, corresponding to wavelengths λ_1 and λ_2, defined as

$$R = \frac{\lambda}{\Delta\lambda} \qquad \Delta\lambda = |\lambda_1 - \lambda_2| \qquad \lambda = \frac{\lambda_1 + \lambda_2}{2}$$

is equal to mN

5.1.3 Fraunhofer diffraction from a circular aperture

A parallel beam of monochromatic light of wavelength λ is incident on a circular aperture (Fig. 5.3). To a very small section dS of the aperture is related, at angle θ, a very small portion $df(\theta)$ of the superposed wave reaching the screen where the diffraction pattern appears. With adS the

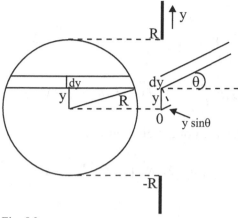

Fig. 5.3

magnitude of the wave (because dS is dimensionally an area, the constant a is necessary to define adS as a differential value of an electric field) and φ the phase difference due to the path difference $y\sin\theta$, we have

$$d\,f(\theta)=a\,dS\,e^{i\varphi}=adS\,e^{iyk\sin\theta} \qquad \varphi=ky\sin\theta$$

The small section is

$$dS=2\sqrt{R^2-y^2}\;dy$$

hence

$$df=ae^{iky\sin\theta}\,2\sqrt{R^2-y^2}\;dy$$

The integral of all the small df due to all small dS from $-R$ to $+R$ is

$$f(\theta)=2a\int_{-R}^{R}e^{iky\sin\theta}\sqrt{R^2-y^2}\,dy$$

Assuming

$$s=\frac{y}{R}\qquad dy=Rds\qquad -R\le y\le R \;\rightarrow\; -1\le s\le 1$$

we have

$$f(\theta)=2a\int_{-1}^{+1}e^{\,i\,Rsk\sin\theta}\ (R\sqrt{1-\frac{y^2}{R^2}}\,)Rds=$$

$$=2aR^2\int_{-1}^{+1}e^{\,i\,g(\theta)\,s}\sqrt{1-s^2}\ ds\qquad g(\theta)=Rk\sin\theta$$

or with $S=\pi R^2$ the area of the circular aperture

$$f(\theta)=\frac{\pi}{\pi}2aR^2\int_{-1}^{+1}e^{\,i\,g(\theta)\,s}\sqrt{1-s^2}\ ds$$

$$=\frac{2}{\pi}\,aS\int_{-1}^{+1}e^{\,i\,g(\theta)\,s}\sqrt{1-s^2}\ ds\ =aSF(\theta)$$

where aS is now the full magnitude of the electric field wave, with

$$F(\theta)=\frac{2}{\pi}\int_{-1}^{+1}e^{\,i\,g(\theta)\,s}\sqrt{1-s^2}\ ds$$

The complex conjugate of $f(\theta)$ is

$$f(\theta)^*=aS\,F^*(\theta)\qquad F^*(\theta)=\frac{2}{\pi}\int_{-1}^{+1}e^{\,-i\,g(\theta)\,s}\sqrt{1-s^2}\ ds$$

The intensity and the corresponding relative intensity are

$$I(\theta)=a^2S^2F(\theta)F^*(\theta)=I_0F(\theta)F^*(\theta)\qquad I_r(\theta)=F(\theta)F^*(\theta)$$

Finding a value of I_r for an assigned θ (with fixed values of R and λ), or the set of their values for the necessary interval of θ that allows a representation of the central maximum and two lateral maxima is a simple task using MATLAB (Fig. 5.4).

For the condition of the first minimum we are looking for a formula similar to that of a single slit

$$\frac{\pi}{\lambda}h\sin\theta=m\pi\quad\rightarrow\quad\frac{\pi}{\lambda}D\sin\theta=m\pi\qquad m=\frac{D\sin\theta}{\lambda}$$

For the angular position θ_{min} of the first minimum (with the same MAT-

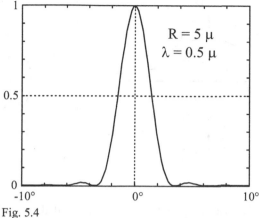

Fig. 5.4

LAB script) we obtain

$$m = \frac{D \sin \theta_{min}}{\lambda} = 1.2195$$

Because the value of θ_{min} is small we can write

$$\frac{D \theta_{min}}{\lambda} = 1.2195$$

Hence the diameter of the circular aperture D and the angular position θ_{min} of the first minimum are inversely proportional.

The relative intensity of the first secondary maximum has a value of about 2% of that of the central maximum. So only the central maximum has significant interest and corresponds to a bright ("Airy") disk surrounded by circular fringes barely observable.

The image in focus region of a far point object given by a lens of an optical instrument (by our eyes, too) is really an Airy diffraction disk. Images corresponding to two nearby far object points are really two Airy disks. These can be considered "resolved" (or distinctly visible) if the distance, between their centers, is greater than their radii. This condition is known as "Rayleigh's criterion".

5.1.4 Fresnel diffraction: the zones and the integrals

A point source O is distant a from a small circular aperture in the plane

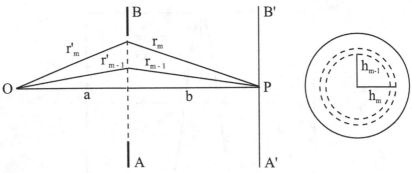

Fig. 5.5

A-B (Fig. 5.5). The rays r'_m and r'_{m-1} diffracted at the height h_m and h_{m-1} in A-B reach the point P of a screen A'-B', distant b from A-B. We call them r_m and r_{m-1} after diffraction. We have, remembering the power series $(1+x)^p$ and that h_m and h_{m-1} are small quantities and using the first two terms of the series,

$$r'_m + r_m = \sqrt{a^2 + h_m^2} + \sqrt{b^2 + h_m^2} = a + b + \frac{G}{2}h_m^2 \qquad G = \frac{a+b}{ab}$$

Fresnel considers the aperture as a set of small annular surfaces, called zones, whose thickness is $h_m - h_{m-1}$, satisfying the condition

$$r'_m + r_m = a + b + m\frac{\lambda}{2}$$

From the previous two formulae we have

$$a + b + \frac{G}{2}h_m^2 = a + b + m\frac{\lambda}{2} \qquad h_m = \sqrt{\frac{m\lambda}{G}} = \sqrt{m}\sqrt{\frac{\lambda}{G}}$$

and the area A_z of a zone is

$$A_z = \pi(h_m^2 - h_{m-1}^2) = \pi\left(\frac{m\lambda}{G} - \frac{(m-1)\lambda}{G}\right) = \frac{\pi\lambda}{G}$$

The heights h_m are proportional to the square root of the number m of the m-zone and all zones have the same area. The "mean" phase changes by π from one zone to the next. The sum of the areas A_z of the zones is the area A_0 of the circular aperture that is small. So the area of each zone is very small. The amplitude E_0 of the electric field on the aperture, proportional to the area of its surface, can be considerer equal to the sum of amplitudes E_{z1}, E_{z2} ... E_{zN} associated to each of the N zones. Therefore for the electric fields E_{z1}, E_{z2} ... E_{zN} the phase changes by π from one to the subsequent zone. Hence the resultant E_P of these fields, when they reach the point P, is only E_{z1}, if N is an odd number. With this assumption we can write for the relative intensity of the point P

$$I_{rP} = \frac{I_p}{I_0} = (\frac{E_P}{E_0})^2 = (\frac{A_z}{A_0})^2 = (\frac{A_z}{NA_z})^2 = \frac{1}{N^2}$$

Greater is N, smaller is I_{rP}. The number N of zones is defined by the formula

$$h_m = \sqrt{m}\sqrt{\frac{\lambda}{G}} \rightarrow R = \sqrt{N}\sqrt{\frac{\lambda}{G}} \qquad N = \frac{R^2 G}{\lambda}$$

where R is the radius of the circular aperture. Hence we have

$$I_{rP} = \frac{1}{N^2} = (\frac{\lambda}{R^2 G})^2$$

The intensity I_{rp} is a complicated function of R, λ, and G (dependent on a and b). Therefore the calculus of the relative intensity on each point P of the screen is a burdensome task and consequently is also difficult to normalize the function I_{rp}. This effort is greatly simplified when the Fraunhofer conditions can be applied.

The general form for the diffraction, valid both in the Fresnel and Fraunhofer condition and known as the Fresnel-Kirchhoff equation, on the point of the screen P of ordinate y and distant r from the aperture is

$$E(y) = D \int_A e^{ikr} dA \tag{5.1}$$

under the assumption that the obliquity and radial factors are constants that can be included in D.

If the aperture is a linear slit (Fig. 5.6) r is

$$r = \sqrt{b^2 + (y-t)^2} = b\left[1 + (\frac{y-t}{b})^2\right]^{\frac{1}{2}}$$

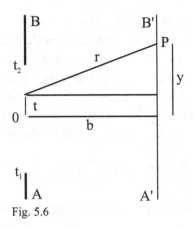

Fig. 5.6

Assuming

$$\frac{y-t}{b} << 1 \tag{5.2}$$

and using the first two terms of the power series, the formula of r can be written

$$r = b + \frac{1}{2}\frac{(y-t)^2}{b} = b + \frac{y^2 - 2yt + t^2}{2b} \tag{5.3}$$

The assumption (5.2) must be fulfilled either within the Fraunhofer than in Fresnel hypothesis.

For the first case the screen A'-B' is assumed far from A-B. Therefore because is t << b, t^2 can be neglected and the formula for r becomes

$$r = b + \frac{y^2 - 2yt}{2b} = b + \frac{y^2}{2b} - \frac{yt}{b} = D_1 - \frac{yt}{b}$$

With this value of r and including also the constant $D1$ in D the formula (5.1) gives the same result as the $sinc(\alpha)$ function (see Sec. 5.1.2).

With small values of b the term t^2 cannot be neglected and the formula (5.3)

$$r = b + \frac{1}{2}\frac{(y-t)^2}{b}$$

must be used for r. Including in D also the constant factor $exp(ikb)$ the formula for integration becomes (hypothesis Fresnel)

$$E(y) = D \int_A e^{ikr} dA = D \int_{t_1}^{t_2} e^{ik\left[b + \frac{1}{2}\frac{(y-t)^2}{b}\right]} dt = D \int_{t_1}^{t_2} e^{i\pi \frac{(y-t)^2}{\lambda b}} dt \quad (5.4)$$

Introducing the new variable

$$\frac{s^2}{2} = \frac{(y-t)^2}{b\lambda} \qquad s = \sqrt{\frac{2}{b\lambda}}(y-t) \qquad ds = -\sqrt{\frac{2}{b\lambda}} dt \qquad ds = Bdt$$

Including too the constant B in D we have the formula of the Fresnel integrals

$$E(y) = D \int_{s_1}^{s_2} e^{i\frac{\pi}{2}s^2} ds = D\left[\int_{s_1}^{s_2} \cos(i\frac{\pi}{2}s^2)ds + i\int_{s_1}^{s_2} \sin(i\frac{\pi}{2}s^2)ds\right] =$$

$$= D[C(s) + iS(s)]_{s_1}^{s_2} \qquad s_1 = \sqrt{\frac{2}{b\lambda}}(y-t_1) \qquad s_2 = \sqrt{\frac{2}{b\lambda}}(y-t_2) \quad (5.5)$$

$E(y)$ can be determined calculating the last two Fresnel integrals using either tables of integral to find $C(s)$ and $S(s)$ and with the aid of the Cornu's spiral or directly the formula (5.4). The alculation of the complex integral can be accomplished with MATLAB. The intensity will be, as usual, given in both cases by the product of $(E(y)$ by its complex conjugate $E(y)*$.

5.2 Problems

5.2.1 Single slit

A beam of *monochromatic* light ($\lambda = 0.5$ µ) is incident normally on the plane A-B where is a single linear slit of width $d/2 = 0.05$ mm. We assume the wave magnitude of the electric field, reaching the slit, has a constant and unitary value. The diffraction pattern on the screen A'-B' is given by the square of the sinc(α) function (Fig. 5.7). Find this result using Fourier series and transform (see Appendix 1).

Solution:

The "top-hat" function (Fig. 5.8) of the unitary electric field can be re-

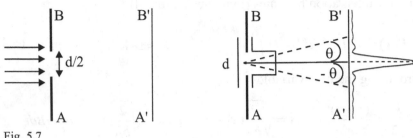

Fig. 5.7

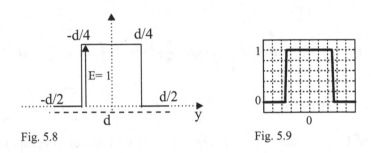

Fig. 5.8

Fig. 5.9

presented, using the Fourier's theorem, by adding together some number of cosine waves of different wavelengths. The requirement of periodicity is not necessary if the periodic function has to be synthesized in a limited interval. The Fourier series for our function is

$$E(y) = \frac{E_0}{2} + \sum_{m=1}^{\infty} E_m \cos(mwy) \qquad w = \frac{2\pi}{d}$$

where y is defined in the interval $[-(d/2), (d/2)]$ and E_m is given by the following formula

$$E_m = \frac{1}{d/2} \int_{-\frac{d}{2}}^{\frac{d}{2}} E(y) \cos(mwy) dy$$

For the function $E(y)$ equal to one in the interval $[-(d/4), (d/4)]$ and equal to zero outside this interval, the preceding formula becomes

$$E_m = \frac{2}{d} \int_{-\frac{d}{4}}^{\frac{d}{4}} \cos(mwy) dy = \frac{2}{d} \left[\frac{\sin(mwy)}{mw} \right]_{-\frac{d}{4}}^{\frac{d}{4}} = \frac{\sin(m\frac{\pi}{2})}{m\frac{\pi}{2}}$$

with E_m not equal to zero for $m = 0$ and for odd values of m. Therefore E_m is defined as a sinc(α) function. Using p for the odd values of m, $p = 1, 3$ ... N we have

$$E(y) = \frac{1}{2} + \sum_{p=1}^{N} E_p \cos(pky) \qquad E_p = \frac{\sin(p\frac{\pi}{2})}{p\frac{\pi}{2}}$$

With $N = 3999$ the "top-hat" function is easily retrieved (Fig. 5.9). The corresponding Fourier transform of the previous symmetric function becomes

$$G(t) = \frac{1}{\pi} \int_0^\infty E(y) \cos(ty) dy \qquad t = \frac{2\pi}{\lambda} \sin\theta$$

with t function of θ (Fig. 5.10).Assuming a finite number of the values of y in the interval $[-(d/4), (d/4)]$ where $E(y)$ is different from zero, the function $G(t)$ can be calculated for a point t.

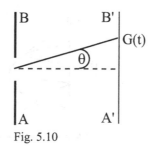

Fig. 5.10

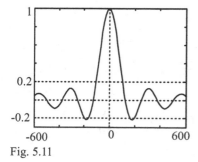

Fig. 5.11

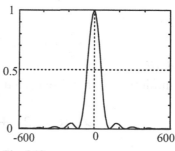

Fig. 5.12

The function $G(t)$ can be complex. The product $G(t)G^*(t)$, where $G^*(t)$ is

the complex conjugate of $G(t)$, is intensity, measured as watt/m^2. The Fourier transform $G(t)$ and product $G(t)G^*(t)$, with MATLAB, are calculated and plotted, as functions of t with θ defined in the interval $(-3,3)°$ (Fig. 5.11 and Fig. 5.12).
The result is the same as that obtained with the classical formula

$$I_r = \sin c^2 = \frac{\sin^2(\alpha)}{\alpha^2} \qquad \alpha = \frac{d}{4}t = \frac{d}{4}\frac{2\pi}{\lambda}\sin\theta$$

5.2.2 Young 1
A beam of monochromatic light ($\lambda = 0.5 \ \mu$) is incident normally on the plane A-B of a Young's apparatus. The slits are d = 0.1 mm apart and h = 0.04 mm wide. The screen A'-B' is distant b = 1 m from A-B (Fig. 5.13). Verify that there are five fringes between the two minima of diffraction of order m = 1. Find the relative intensity of the fringe of the second order of interference.

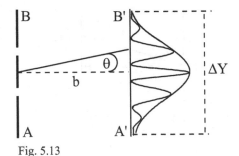

Fig. 5.13

Solution:
A minimum of interference requires

$$d\sin\theta = (2m+1)\frac{\lambda}{2} \qquad \theta = (2m+1)\frac{\lambda}{2d}$$

$$y_m = b\tan\theta = b\theta = b\,(2m+1)\frac{\lambda}{2d}$$

and for $m^* = m + 1$

$$y_{m^*} = b(2m^*+1)\frac{\lambda}{2d}$$

The width of a fringe is

$$\delta y = y_{m^*} - y_m = \frac{b\lambda}{d} = \frac{10^3 \cdot 5 \cdot 10^{-4}}{0.1} = 5\,\text{mm}$$

The minimum of diffraction, for $m = 1$, requires

$$h\sin\theta = \lambda \qquad \theta = \sin\theta = \frac{\lambda}{h}$$

The distance of a minimum of order $m = 1$ from the center of the maximum of intensity is

$$Y = b\theta = \frac{b\lambda}{h}$$

Between the two minima of order $m = 1$ the distance Δy is

$$\Delta Y = 2\frac{b\lambda}{h} = 25\,\text{mm}$$

and the number of fringe is

$$\frac{\Delta Y}{\Delta y} = \frac{2b\lambda}{h}\frac{d}{b\lambda} = \frac{2d}{h} = \frac{2 \cdot 0.1}{0.04} = 5$$

For the fringe of interference of the second order ($m = 2$) we have

$$\sin\theta = \frac{2\lambda}{d} = \frac{2 \cdot 5 \cdot 10^{-4}}{0.1} = 0.01$$

$$\alpha = \frac{h\pi}{\lambda}\sin\theta = \frac{h\pi}{\lambda}\frac{2\lambda}{d} = 2\pi\frac{h}{d} = 144.0°$$

$$\beta = \frac{d\pi}{\lambda}sen\theta = \frac{d\pi}{\lambda}\frac{2\lambda}{d} = 2\pi = 360°$$

Hence the requested relative intensity is

$$I_r = (\frac{\sin\alpha}{\alpha})^2 \cos^2\beta = (\frac{\sin 2.51}{2.51})^2 = 0.06$$

5.2.3 Young 2

A beam of monochromatic light ($\lambda = 0.5$ μ) is incident normally on the plane A-B of a Young's apparatus. The slits are $d = 0.1$ mm apart. The screen A'-B' is distant $b = 1$ m from A-B and the bright fringe of order $m = 1$ is distant y from the origin 0 (Fig. 5.14). If a thin transparent film of refractive index n and thickness $s = 0.02$ mm covers the lower slit this bright fringe is shifted below of 13 mm. The bright fringe has now a distance y^* from the origin 0. Also the bright fringe of order $m = 0$ assume a new position y_0 from the origin 0. Find the heights y, y^*, y_0 and the value of the refractive index n.

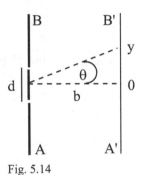

Fig. 5.14

Solution:

When the film is absent for the bright fringe of order $m = 1$ we have

$$d\sin\theta = \lambda \quad \sin\theta = \frac{\lambda}{d} \quad \theta = \frac{\lambda}{d} \quad y = b\tan\theta = b\theta = b\frac{\lambda}{d} = 5\,\text{mm}$$

Hence the new position will be $y^* = y - 13$ mm = - 8 mm (Fig. 5.15).
For the presence of the film we must have for the new position of the bright fringe of order $m = 1$

$$d\sin\theta^* + s(n-1) = \lambda \quad \sin\theta^* = \frac{\lambda}{d} - \frac{s(n-1)}{d} = \sin\theta - \frac{s(n-1)}{d}$$

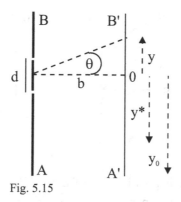

Fig. 5.15

Hence

$$\sin\theta^* = \tan\theta^* = \frac{y - \Delta y}{b} = \frac{\lambda}{d} - \frac{s(n-1)}{d} \qquad \Delta y = b\frac{s(n-1)}{d}$$

remembering that $y = b\lambda/d$. Then

$$n = 1 + \frac{d\,\Delta y}{b\,s} = 1 + \frac{0.1 \cdot 13}{1000 \cdot 0.002} = 1.65$$

For the fringe of order $m = 0$ we have

$$d\sin\theta_0 = -s(n-1) \qquad \theta_0 = -\frac{s(n-1)}{d}$$

and, as expected,

$$y_0 = b\theta_0 = -\frac{bs(n-1)}{d} = -13\,\text{mm}$$

5.2.4 Young 3

Two beams of monochromatic light ($\lambda = 0.5\ \mu$) S and S' are incident on the plane A-B of a Young's apparatus. The first beam S is incident normally and S' with an angle α on A-B (Fig. 5.16). When the slits are $d = d_1 = 100\ \mu$ apart the fringes disappear. Increasing the distance between the slits from d_1 to $d = d_2 = 113\ \mu$, the fringes again disappear. Find the order m^* of the dark fringe due to the source S' superposed to the central bright fringe of order $m = 0$ (angular position $\theta = 0$) due to the source S,

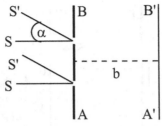

Fig. 5.16

the angle α, the angular position θ^* of the bright fringe of order $m = 0$ due to the source S' that superposes to the dark fringe of order m° due to the source S.

Solution:

The conditions for maxima and minima for source S are (Fig. 5.17)

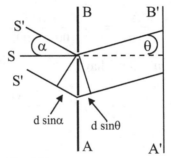

Fig. 5.17

$$d \sin \theta = m \lambda \qquad d \sin \theta = (2m + 1)\frac{\lambda}{2}$$

and for the source S'

$$d (\sin \alpha + \sin \theta) = m \lambda \qquad d(\sin \alpha + \sin \theta) = (2m + 1)\frac{\lambda}{2}$$

For $d = d_1$ there is a maximum for S of order $m = 0$ at the angular position $\theta = 0$. In the same position (Fig. 5.18) must be a minimum for S' of order m^*. Hence

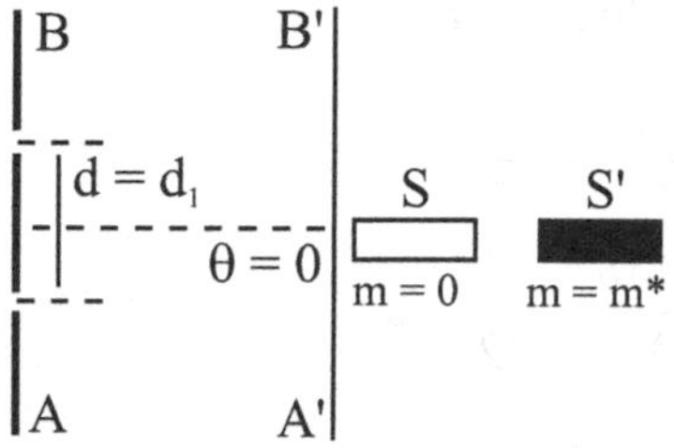

Fig. 5.18 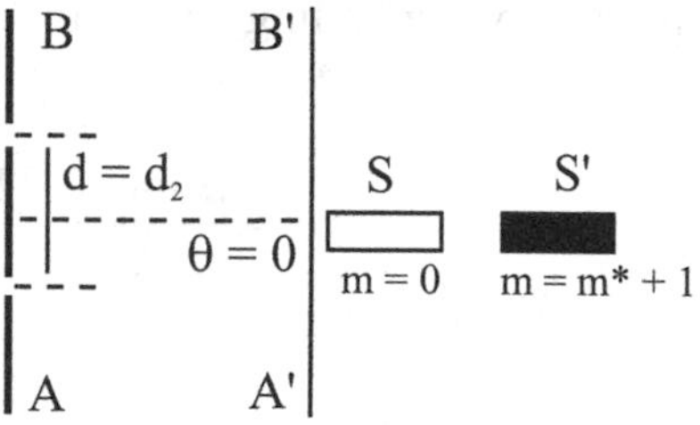   Fig. 5.19

$$d_1 \sin\alpha = (2m^* + 1)\frac{\lambda}{2}$$

For $d = d_2$ there is again a maximum for S of order $m = 0$ at angular position $\theta = 0$ and the same position (Fig. 5.19) must be a minimum for S' of order $m^* + 1$. Hence

$$d_2 \sin\alpha = [2(m^* + 1) + 1]\frac{\lambda}{2} \qquad d_2 \sin\alpha = (2m^* + 3)\frac{\lambda}{2}$$

Dividing the two results we have

$$m^* = \frac{3d_1 - d_2}{2(d_2 - d_1)} = 7$$

For the source S' there is a condition of minimum for $\theta = 0$ and $m^* = 7$. Hence

$$d_1 \sin\alpha = (2m^* + 1)\frac{\lambda}{2} = 15\frac{\lambda}{2} \qquad \alpha = \arcsin\frac{15 \cdot 0.5}{2 \cdot 100} = 2.15°$$

The angular position of the bright fringe, due to the source S', of order $m = 0$ is given by the condition

$$d(\sin\alpha + \sin\theta) = m\lambda \qquad \sin\alpha = -\sin\theta^* \qquad \theta^* = -\alpha = -2.15°$$

For the dark fringe, of angular position θ^* due to the source S, the minimum condition is (Fig. 5.20)

$$d \sin\theta^* = (2m° + 1)\frac{\lambda}{2}$$

and the corresponding orders for the two values of d are

$$m° = \frac{d\sin\theta^*}{\lambda} - \frac{1}{2} = \begin{cases} \dfrac{d_1\sin\theta^*}{\lambda} - \dfrac{1}{2} = 7 \\ \dfrac{d_2\sin\theta^*}{\lambda} - \dfrac{1}{2} = 8 \end{cases}$$

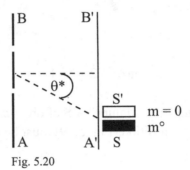

Fig. 5.20

5.2.5 *Young 4*

A beam of monochromatic light ($\lambda = 0.5\ \mu$) is incident on the plane A-B of a Young's apparatus. The slits are $d = 0.1$ mm apart and have a width $h = 0.01$ mm. The fringes are observed on the screen A'-B' distant $b = 1$ m from A-B. Find the order m and the corresponding position y of the first bright fringe whose relative intensity is less than 30% of the maximum of the relative intensity (Fig. 5.21).

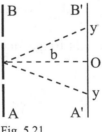

Fig. 5.21

Solution:

The relative intensity of a bright fringe is

$$I_r = \frac{\sin^2 \alpha}{\alpha^2} \qquad \alpha = \pi \frac{h}{\lambda} \sin(\theta)$$

The condition for the maxima of interference is

$$d \sin \theta = m\lambda \qquad \sin \theta = \frac{m\lambda}{d}$$

Hence

$$\alpha = \frac{m\pi h}{d} \qquad m = 0, 1, 2, ., ., .$$

Plotting I_r as function of m (Fig. 5.22) we find $I_r = 0.25$ for $m = 6$. In Table 5.1 we see that the bright observable fringes are about seven.

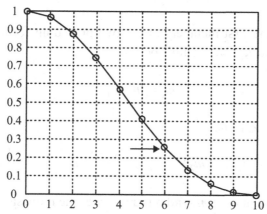

Fig. 5.22 Relative intensity function of m

Table 5.1

m	0	1	2	3	4	5	6	7	8	9	10
I_r	1.00	0.97	0.88	0.74	0.57	0.41	0.25	0.14	0.05	0.01	0.00

For $m = 6$ we have

$$\theta = \pm \arcsin(\frac{m\lambda}{d}) = 1.7° \qquad y = \pm b \arctan \theta = \pm 30\,\text{mm}$$

5.2.6 Young 5

Two beams of monochromatic light ($\lambda_1 = 0.4\ \mu$ and $\lambda_2 = 0.6\ \mu$) are incident on the plane A-A' of a Young's apparatus. Find the angular positions where the bright fringes of the two wavelengths are superposed and the corresponding orders.

Solution:

The conditions for the maxima for the two wavelengths are

$$d \sin \theta_1 = m_1 \lambda_1 \qquad d \sin \theta_2 = m_2 \lambda_2$$

If the fringes are superposed it is

$$\theta_1 = \theta_2$$

and, because the first members of the two preceding formulae are supposed equal and $\lambda_1 < \lambda_2$, it must be

$$m_1 = m_2 + p$$

with p an integer. Hence

$$(m_2 + p)\lambda_1 = m_2 \lambda_2 \qquad p = \frac{\lambda_2 - \lambda_1}{\lambda_1} m_2 \qquad p = \frac{1}{2} m_2$$

Remembering that $m_1 > m_2$ and p is an integer the requested values of m_2 must be even numbers. The results are in (Table 5.2)

Table 5.2

m_2	2	4	6
p	1	2	3
$m_1 = m_2 + p$	3	6	9

The corresponding angular positions (Fig. 5.23) are obtained varying m from 0 to 10 in the formulae

$$\theta_1 = \arcsin(\frac{m\lambda_1}{d}) \qquad \theta_2 = \arcsin(\frac{m\lambda_2}{d})$$

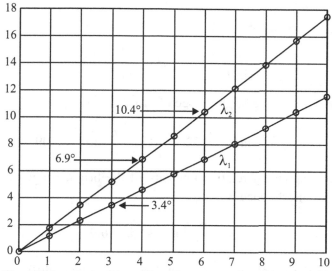

Fig. 5.23 Angular positions (degree) versus m (fringe order)

5.2.7 *Young 6*

A beam of monochromatic light ($\lambda = 0.5\ \mu$) is incident on the plane A-B of a Young's apparatus to which a partial plot of fringes is added. The angular position of the bright fringe due to interference of the first order is $\theta_1 = 0.94°$ and the angular position of the first dark fringe due to the diffraction is $\theta_2 = 5.7°$ (Fig. 5.24). Find the relative intensity corresponding to these angular positions and the values of h (width of the slits) and d (distance between them).

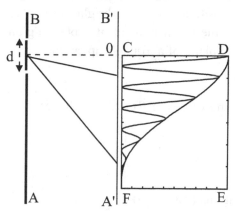

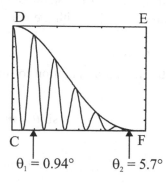

Fig. 5.24

Solution:
For $m = 1$ the values of h and d are

$$h = \frac{\lambda}{\sin\theta_1} = 5\mu \qquad d = \frac{\lambda}{\sin\theta_2} = 30\mu$$

Hence we have for the bright fringe of order $m = 1$

$$\alpha_1 = \frac{h\pi}{\lambda}\sin\theta_1 = 0.52 \qquad \beta_1 = \frac{d\pi}{\lambda}\sin\theta_1 = 3.09$$

and the corresponding relative intensity of the bright fringe of the first order due to interference is

$$I_{r1} = (\frac{\sin\alpha_1}{\alpha_1})^2 \cos^2\beta_1 = 0.9$$

the corresponding values for the first dark fringe due to diffraction are

$$\alpha_2 = \frac{h\pi}{\lambda}\sin\theta_2 = 3.12 \qquad \beta_2 = \frac{d\pi}{\lambda}\sin\theta_2 = 18.7$$

$$I_{r2} = (\frac{\sin\alpha_2}{\alpha_2})^2 \cos^2\beta_2 = 10^{-4}$$

5.2.8 Young 7
A beam of monochromatic light ($\lambda = 0.51$ μ) is incident normally on the plane of a Young's apparatus with a distance d $= = 1$ μ between the slits. Find the orders m and the angular positions of all the bright observable fringes. Find the same values when the beam is incident subtending an angle $\alpha = 30°$ with the normal to the plane of a Young's apparatus.

Solution:
In the first assumption for the extreme values of θ ($\pm 90°$) we have

$$d\sin\theta = m\lambda \qquad m = \pm\frac{d}{\lambda} = \pm 1.2 = \begin{cases} 1 \\ 0 \\ -1 \end{cases}$$

and the corresponding angular positions (Fig. 5.25)

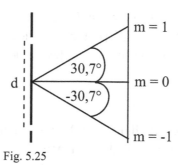

Fig. 5.25

$$\theta = 0° \qquad \theta = \pm 30.7°$$

Indeed if we put $m = 2$ we would have

$$\theta = \arcsin(\frac{2\lambda}{d}) = (90 + i11.4)°$$

In the second assumption the condition for the maxima becomes (Fig. 5.26)

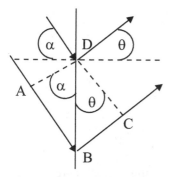

Fig. 5.26

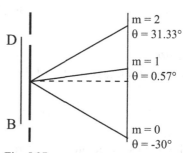

Fig. 5.27

$$d(\sin\alpha + \sin\theta) = m\lambda$$

and for the extreme values of θ ($\pm 90°$) we have

$$d(\sin\alpha \pm 1) = m\lambda \qquad m = \begin{cases} (\sin\alpha + 1)\dfrac{d}{\lambda} = 2 \\ (\sin\alpha - 1)\dfrac{d}{\lambda} = 0 \end{cases}$$

Hence for the allowable value for m 0, 1, 2, the corresponding angular positions are -30°, 0.57°, 31.33° (Fig. 5.27). Indeed if we assigned to m the values -1 or 3 we would have complex values for the angular positions.

5.2.9 Young 8
A beam of monochromatic light ($\lambda = 0.7$ μ) is incident normally on the plane of a Young's apparatus with a distance $d = 5\lambda$ between the slits that are large $h = 3\lambda$. A thin transparent film, of refractive index $n = 1.38$, covering the higher slit, produces a phase shift of π (Fig. 5.28). Find the thickness s of the film and the angular position of the fringes of maximum intensity.

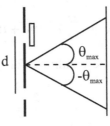

Fig. 5.28

Solution:
For the thickness of the film we have

$$s(n-1) = \frac{\lambda}{2} \qquad s = \frac{\lambda}{2(n-1)} = 0.92\,\mu$$

The waves associated to the ray emerging from the two slits are

$$y_1 = ae^{i(\alpha+\pi)} \qquad y_2 = ae^{i(\alpha+\varphi)} \qquad \varphi = kd\sin\theta$$

and their sum is

$$y = ae^{i(\alpha+\pi)} + ae^{i(\alpha+\varphi)} = a(e^{i\alpha}e^{i\pi} + e^{i\alpha}e^{i\varphi}) = ae^{i\alpha}(-1+e^{i\varphi})$$

The intensity is obtained multiplying y with its complex conjugate

$$I_{r2} = a^2(-1+e^{i\varphi})(-1+e^{-i\varphi}) = 2a^2(1-\cos\varphi)$$

that, with

$$\beta = \frac{\varphi}{2} = \frac{\pi}{\lambda} d \sin\theta$$

and a usual trigonometric formula, becomes

$$I_2 = 4a^2 \sin^2\beta \qquad I_{r2} = \frac{I_2}{4a^2} = \sin^2\beta$$

The relative intensity due to the diffraction is

$$I_{r1} = \left(\frac{\sin\alpha}{\alpha}\right)^2 \qquad \alpha = \frac{\pi}{\lambda} h \sin\theta$$

Hence the real relative intensity is

$$I_r = I_{r1}I_{r2} = \frac{\sin^2\alpha}{\alpha^2}\sin^2\beta$$

with α and β functions of θ. For $\theta = 0$ we have $I_{r1} = 1$ but $I_{r2} = 0$. Therefore for $\theta = 0$ is $I_r = 0$, too. The angular positions $\theta = \pm 5.2°$ for the maxima of intensity ($I_r = 0.76$) are obtained plotting the relative intensity I_r versus θ defined in the interval $\pm 20°$ (Fig. 5.29). There are other two

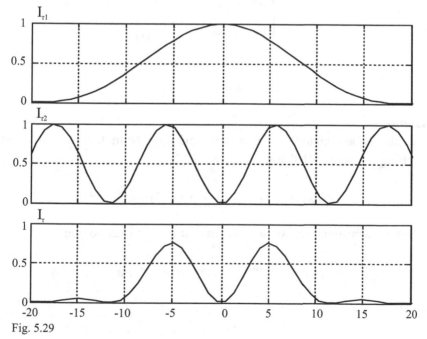

Fig. 5.29

angular positions ($\theta = \pm17.5°$) of barely observable bright fringes ($I_r = 0.01$).

5.2.10 Young 9

A beam of monochromatic light ($\lambda = 0.5\ \mu$) is incident normally on the screen A-B of a Young's apparatus with a distance $d = 0.05$ mm between the slits. A thin transparent film, of refractive index $n_1 = 1.78$, cover the higher slit, and another film, of refractive index $n_1 = 1.42$, covers the lower slit (Fig. 5.30). Both have the same thickness $s = 0.01$ mm. Find the vertical positions y (in mm), the angular positions θ (in degree) and the orders m of the observable bright fringes.

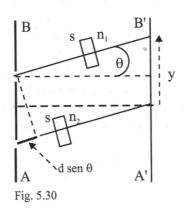

Fig. 5.30

Solution:

The phases associated to the waves emerging from the two slits and reaching the screen BB' are

$$\alpha_1 = ks(n_1 - 1) \qquad \alpha_2 = ks(n_2 - 1) + kd\sin\theta$$

with the corresponding wave function and its complex conjugate

$$y = a(e^{i\alpha_1} + e^{i\alpha_2}) \qquad y^* = a(e^{-i\alpha_1} + e^{-i\alpha_2})$$

The intensity is

$$I = yy^* = a^2 \left[(e^{i\alpha_1} + e^{i\alpha_2})(e^{-i\alpha_1} + e^{-i\alpha_2}) \right] =$$

$$= a^2 \left[e^{i(\alpha_1 - \alpha_2)} + e^{-i(\alpha_1 - \alpha_2)} \right] =$$

$$= a^2(2 + e^{i\gamma} + e^{-i\gamma}) = a^2(2 + 2\cos\gamma) = 2a^2(1 + \cos^2\frac{\gamma}{2} - \sin^2\frac{\gamma}{2})$$

Hence

$$I = 4a^2 \cos^2\frac{\gamma}{2} \qquad I_{r1} = \frac{I}{4a^2} = \cos^2\frac{\gamma}{2}$$

with

$$\gamma = \alpha_1 - \alpha_2 = ks(n_1 - 1) - [ks(n_2 - 1) + kd\sin\theta] = k[s(n_1 - n_2) - d\sin\theta] =$$

$$= k[s(n_1 - n_2) - d\sin\theta]$$

The condition for the maximum of I_{r1} is

$$\frac{\gamma}{2} = m\pi \qquad m = 0,1,2,...$$

or

$$\frac{k}{2}[s(n_1 - n_2) - d\sin\theta] = m\pi \qquad [s(n_1 - n_2) - d\sin\theta] = m\lambda$$

$$d\sin\theta = s(n_1 - n_2) - m\lambda \qquad \theta = \arcsin(\frac{s(n_1 - n_2) - m\lambda}{d})$$

The relative intensity due to the diffraction and the real relative intensity are (Fig. 5.31)

$$I_{r2} = (\frac{\sin\alpha}{\alpha})^2 \qquad \alpha = \pi\frac{h}{\lambda}\sin\theta \qquad I = I_{r2}I_{r1} = (\frac{\sin\alpha}{\alpha})^2 \cos^2\frac{\gamma}{2}$$

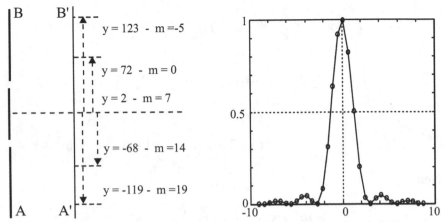

Fig. 5.31 Vertical positions (in mm) of the five maxima and their orders (left). Intensity versus angular position θ (in degree, right)

The numerical data are in Table 5.3

Table 5.3

m	-5	0	7	14	19
θ (°)	7.0	4.1	0.1	-3.9	-6.8
y (mm)	122.9	72.2	2.0	-68.2	-118.8
I_r		0.02	0.05	1.0	0.05	0.01

5.2.11 A five-slits grating

A beam of *monochromatic* light ($\lambda = 0.5\ \mu$) is incident normally on a grating formed by five slits. The distance between the slits is $d = 0.01$ mm and the width of each slit is $h = 0.005$ mm. We assume the magnitude of wave of the electric field entering the slits has a unitary value (Fig. 5.32).

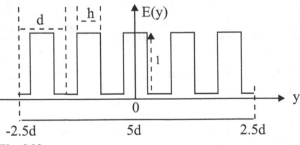

Fig. 5.32

Find the Fourier synthesis (see Appendix 1) of the electric field on the slits, the Fourier transform and the intensity on the screen where the fringes appear in the angular interval (-1.5, 1.5)°. Compare the results with the relative intensity obtained using the standard Fraunhofer formula for $N = 5$.

Solution:

Computing the Fourier series (see Sec. 5.2.1) for the five adjacent "top-hat" function we obtain the numerical values of the series in the array $f(y)$ with y defined in the interval (-2.5d, 2.5d) mm (Fig. 5.33). The corresponding Fourier transform for the angular position θ is numerically determined for a set of values of θ defined in the interval (-1.5, 1.5)° (Fig. 5.34)

$$G(\theta) = \frac{1}{\pi} \int_{-2.5d}^{2.5d} f(y)\cos(ky)dy \qquad k = \frac{2\pi}{\lambda}\sin\theta$$

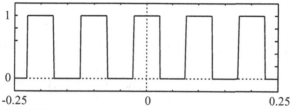

Fig. 5.33 In abscissa y in mm

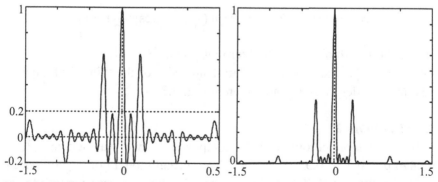

Fig. 5.34 In the two abscissae the angular positions θ (in degree). The Fourier transform (left) and the Spectral Power Density (right)

Normalizing (to a maximum unitary value) $G(\theta)$ and defining its complex conjugate $G^*(\theta)$, the values of the intensity, in the assigned interval, are given by their product. These values are identical with those derived from the formula

$$I_r = (\frac{\sin\alpha}{\alpha})^2 \, (\frac{\sin 5\beta}{5\sin\beta})^2 \qquad \alpha = \pi\frac{h}{\lambda}\sin\theta \qquad \beta = \pi\frac{d}{\lambda}\sin\theta$$

for the Fraunhofer diffraction pattern of the relative intensity when $N = 5$.

5.2.12 Grating 1

A beam of monochromatic light ($\lambda = 0.58958$ μ) is incident normally on a grating composed of $N = 1000$ equidistant parallel and linear slits. The distance between the slits is $d = 8$ μ and each slit is large $h = 2$ μ. Find the angular positions and the relative intensity of the "lines" included between the angular position $0°$ and that corresponding to the first minimum of the diffraction. Verify what happens to the line of the fourth order.

Solution:

The angular position $\theta_{\min}$ of the first minimum due to the diffraction is

$$\theta_{\min} = \arcsin(\frac{\lambda}{h}) = \arcsin(\frac{\lambda}{2}) = 17.1°$$

For the line of the fourth order we have

$$d\sin\theta = m\lambda \qquad \theta_{m=4} = \arcsin(\frac{4\lambda}{8}) = \arcsin(\frac{\lambda}{2}) = 17.1°$$

Hence the fourth order line has zero intensity.
The angular positions and the relative intensity of the "lines" of angular position smaller than $\theta_{\min}$ are given in Fig. 5.35.

5.2.13 Grating 2

A beam of monochromatic light composed of two wavelengths ($\lambda_1 = 0.5890$ μ and $\lambda_2 = 0.5896$ μ) is incident normally on a grating composed of N equidistant parallel and linear slits. The distance between the slits is d. The two lines, both of the fourth order, corresponding to the two wave-

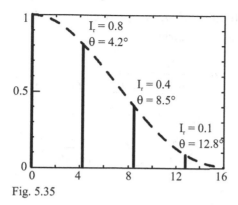

Fig. 5.35

lengths, have angular positions whose difference is $\alpha^* = 4.4'$. Find the values of d and N.

Solution:

From the condition for the maximum of the order m for the first and the second wavelength and using α as the value in radians of α^* we have

$$d \sin \theta_1 = m\lambda_1 \qquad \sin \theta_1 = \frac{m\lambda_1}{d} \qquad d \sin(\theta_1 + \alpha) = m\lambda_2$$

with

$$\cos \theta_1 = \sqrt{1 - \sin^2 \theta_1} = \sqrt{1 - \frac{m^2 \lambda_1^2}{d^2}} = \frac{1}{d}\sqrt{d^2 - m^2 \lambda_1^2}$$

For λ_2 we have

$$d(\sin \theta_1 \cos \alpha + \sin \alpha \cos \theta_1) = m\lambda_2$$

and using the previous formulae for $\sin \theta_1$ and $\cos \theta_1$

$$d\left(\frac{m\lambda_1}{d} \cos \alpha + \sin \alpha \frac{1}{d}\sqrt{d^2 - m^2 \lambda_1^2}\right) = m\lambda_2$$

After some simple algebraic operations, we have for d with $m = 4$

$$d^2 = m^2 \frac{\lambda_1^2 + \lambda_2^2 - 2\lambda_2 \lambda_1 \cos \alpha}{\sin^2 \alpha} = m^2 A \qquad d = 4\sqrt{0.567} = 3\mu$$

Using, for a small value of α, the approximation $\sin\alpha = \alpha$ and $\cos\alpha = 1$ a value for d can also be retrieved

$$d = \frac{m\lambda_1}{\sin\theta_1} = \frac{m\lambda_2}{\sin\theta_2} = \frac{m\lambda_2}{\sin(\theta_1 + \alpha)}$$

and

$$\frac{\lambda_1}{\sin\theta_1} = \frac{\lambda_2}{\sin\theta_1 + \alpha\cos\theta_1}$$

Resolving for θ_1 we have

$$(\lambda_2 - \lambda_1)\sin\theta_1 = \lambda_1\alpha\cos\theta_1 \qquad \theta_1 = \arctan(\frac{\alpha\lambda_1}{\lambda_2 - \lambda_1}) = 51.48°$$

and

$$d = \frac{4\lambda_1}{\sin\theta_1} = 3\mu$$

From the formula of resolving power the value of N is obtained

$$N = \frac{\lambda_1}{4\Delta\lambda} = 245$$

5.2.14 Grating 3

A beam of monochromatic light composed of two wavelengths ($\lambda_1 = 0.58900 \ \mu$ and $\lambda_2 = 0.58958 \ \mu$) is incident normally on a grating composed of N equidistant parallel slits. The distance between the slits is $d = 6 \ \mu$ and each slit is large $h = 2 \ \mu$. The two lines of the first order must be resolved according to the Rayleigh's criterion. Define what is necessary to fulfill this requirement. Then verify that the criterion is satisfied checking that the angular position of the maximum of the first line is equal to the minimum of the second line. Besides verify that the intensity I_{rs}, sum of the intensities of the two lines, at the "saddle" point is $8/\pi^2$ (= 0.81) times the maximum intensity I_{r1} of the first or that I_{r2} of the second line.

Solution:
Clearly the following formula, with $m = 1$, must be used

$$\frac{\lambda}{\Delta\lambda} = mN \qquad N = \frac{\lambda}{m\Delta\lambda} = \frac{\lambda}{\Delta\lambda} = 1016$$

Using the standard formula

$$I_r = \left(\frac{\sin\alpha}{\alpha}\right)^2 \left(\frac{\sin N\beta}{N\sin\beta}\right)^2 \qquad \alpha = \pi\frac{h}{\lambda}\sin\theta \qquad \beta = \pi\frac{d}{\lambda}\sin\theta$$

varying θ the requested intensities for lines of the first order are calculated and plotted (Fig. 5.36). We find

$$\theta_{1\max} = \theta_{2\min} = 5.63° \qquad \theta_{2\max} = \theta_{1\min} = 5.64° \qquad \Delta\theta = 0.3'$$

And for the "saddle" point is verified the expected relation

$$\frac{I_{rs}}{I_{r1}} = \frac{8}{\pi^2} = \frac{0.55}{0.68} = 0.81$$

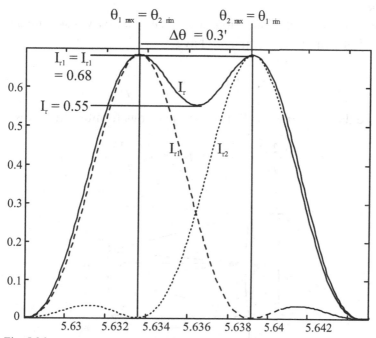

Fig. 5.36

5.2.15 Airy disk 1

A thin lens has diameter $D = 20$ mm and focal length $f = 100$ mm. The point object P has distance z from the lens much greater than the focal length so that the point image P' can be considered in the focal plane of the lens (Fig. 5.37). The object irradiates a monochromatic ($\lambda = 0.5$ μ) light that reaches the lens with an intensity $I_0 = 100$ watt/m². The image cannot be a geometrical point of zero area: it would have an infinite intensity. Find the area A representing the point image P' and the corresponding intensity I. Suppose the lens doesn't absorb energy.

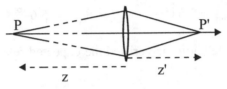

Fig. 5.37

Solution:

The smallest possible dimensions are those of an Airy disk with a radius r (Fig. 5.38) and area $A = \pi r^2$. We have

$$r = f \tan \theta \qquad \sin \theta = \frac{1.22\lambda}{D}$$

Because the angle θ is small, from the previous formulae it follows

$$r = f \frac{1.22\lambda}{D} = 3\mu \qquad A = \pi (f \frac{1.22\lambda}{D})^2 = 29\mu$$

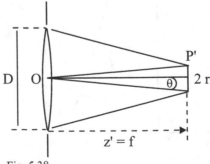

Fig. 5.38

The power reaching the lens

$$P = I_0 \pi (\frac{D}{2})^2$$

reaches also the Airy disk. Hence

$$I = \frac{P}{A} = I_0 \pi (\frac{D}{2})^2 \frac{D^2}{\pi (1.22 f \lambda)^2} = \frac{I_0 D^4}{5.95 f^2 \lambda^2} = 1.1 \cdot 10^9 \frac{\text{watt}}{\text{m}^2}$$

5.2.16 Airy disk 2

A thin lens has diameter $D = 10$ mm and focal length $f = 20$ mm. The point objects A and B are at a distance z from the lens much greater than the focal length so that the point images A' and B' can be considered in the focal plane of the lens (Fig. 5.39). The point objects, distant $y = 0.5$ m from the optic axis, irradiates a monochromatic violet light ($\lambda = 0.4$ μ). Find the maximum distance z_{max} for which, in agreement with the Rayleigh's criterion, the point images A' and B' can see resolved. Find how changes z_{max} if the monochromatic light is red ($\lambda = 0.7$ μ).

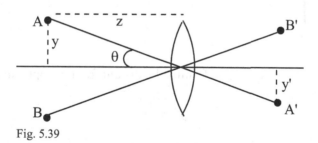

Fig. 5.39

Solution:
From the magnification formula we have

$$y' = y \frac{f}{z}$$

The radius of the Airy disk (Fig. 5.40) is

$$y^* = f \theta^* \qquad \sin \theta^* = \theta^* = \frac{1.22 \lambda}{D} \qquad y^* = f \frac{1.22 \lambda}{D}$$

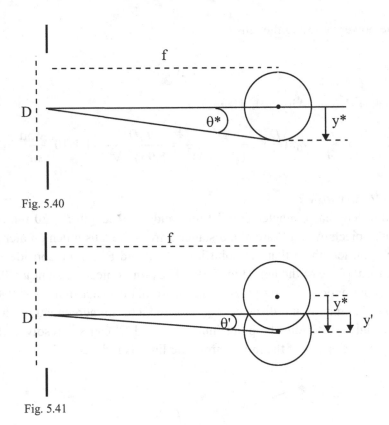

Fig. 5.40

Fig. 5.41

The two point images A' and B' must be distant from the optic axis at least (Fig. 5.41)

$$y' = \frac{y*}{2}$$

according to the Rayleigh's criterion. Hence

$$z_{max} = y\frac{f}{y'} = 2y\frac{f}{y*} = \frac{2yD}{1.22\lambda} = \begin{cases} 11.7\,\text{km} & \text{for red light} \\ 20.5\,\text{km} & \text{for violet light} \end{cases}$$

The maximum distance z_{max} for the two wavelengths is independent from the focal length, increases with y and D and decreases with λ.

5.2.17 Fresnel zones 1

The slit is the area between two concentric circles of the plane A-B: the inner circle is the curve bounding the opaque disk of radius R_1 and the outer one is a circular opening of radius R_2 in the opaque plane A-B (Fig. 5.42). The radii R_1 and R_2 are really a bit less than 2 mm and 4 mm. A beam of monochromatic light ($\lambda = 0.5\ \mu$) is incident on the surface A-B. The source O is distant $a = 700$ mm from A-B. When the distance between A-B and the screen A'-B' is about $b = 360$ mm a spot of intensity different from zero is present in the point P. Evaluate this relative intensity and more accurate values of R_1 and R_2 when the number of zones in the annular area is exactly an odd number N.

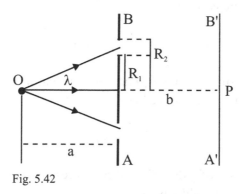

Fig. 5.42

Solution:

The area S of the annular slit is (Fig. 5.43)

$$S = \pi (R_2{}^2 - R_1{}^2) = 37.70\,\text{mm}^2$$

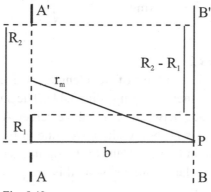

Fig. 5.43

All Fresnel zones (see Sec. 5.1.4) have the same area A_z

$$A_z = \frac{\pi\lambda}{G} \qquad G = \frac{a+b}{ab} \qquad A_z = \frac{\pi\lambda(a+b)}{ab} = 0.37\,\text{mm}^2$$

Hence the number N of the zones in the slit is

$$N = \frac{S}{Az} = 101$$

and the relative intensity is

$$I_r = (\frac{A_z}{S})^2 = \frac{1}{N^2} = 1.0\cdot10^{-4}$$

If only a circular slit of radius R_1 had been present in the plane A-B, the number N_1 of zones would have been

$$N_1 = \frac{R_1 G}{\lambda} = 33.6$$

If only a circular slit of radius R_2 had been present in the plane A-B, the number N_2 of zones would have been

$$N_2 = \frac{R_2 G}{\lambda} = 134.6$$

Rounding N_1 to 33 and N_2 to 134 so that their difference is an odd number $N = 101$, we have the more accurate values R_{1T} and R_{2T} for the radii

$$R_{1T} = \sqrt{\frac{33\lambda}{G}} = 1.98\,\text{mm} \qquad R_{2T} = \sqrt{\frac{134\lambda}{G}} = 3.99\,\text{mm}$$

5.2.18 Fresnel zones 2

A beam of monochromatic light of wavelength λ is incident on the plane A-B where there is a linear opening (Fig. 5.44). The source O is distant a from A-B. The distance between A-B and the screen A'-B' is b. The point P on the screen A'-B' is H_1 below the lower line of the opening. A linear bright fringe, normal in P to the plane of the figure, is present in the plane A'-B'. Verify if the relative intensity of this fringe can be determined using the Fresnel zones.

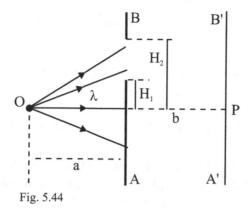

Fig. 5.44

Solution:

We have also in this situation

$$r'_m + r_m = \sqrt{a^2 + h_m^{\,2}} + \sqrt{b^2 + h_m^{\,2}} = a + b + \frac{G}{2}h_m^{\,2} \qquad G = \frac{a+b}{ab}$$

The Fresnel condition is

$$r'_m + r_m = a + b + m\frac{\lambda}{2}$$

Subtracting the previous two formulae the value h_m is obtained (Fig. 5.45)

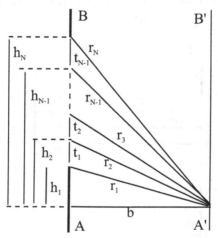

Fig. 5.45

$$h_m = \sqrt{\frac{m\lambda}{G}}$$

Now we have linear zones A_{zm} whose areas are proportional to the thicknesses t_m

$$t_1 = h_2 - h_1 \qquad t_2 = h_3 - h_2 \qquad t_{N-1} = h_N - h_{N-1}$$

The h_m and consequently the t_m are not constant. In Table 5.4 the *first ten* values of h and the *first nine* values of t are given in micron with the initial assumptions are: $\lambda = 0.5$ μ, and (in mm) $H_1 = 2$, $H_2 = 10$, $a = 700$, $b = 500$. The corresponding maximum numbers of h and t are 659 and 658.

Table 5.4

h	382	540	661	764	854	935	1010	1080	1146	1208
t	158	121	102	90	82	75	70	66	62	

The corresponding electric fields are in sequence shifted of π but they have different magnitude and therefore nothing can be known about their superposition when they reach the point P.

5.2.19 Fraunhofer and Fresnel diffraction for a single-slit
On a single-slit is incident, normally to A-B (Fig. 5.46) a parallel beam of monochromatic light ($\lambda = 0.5$ μ).

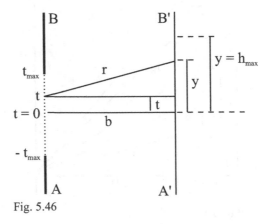

Fig. 5.46

Find a workable set of values of t_{max}, h_{max} and b that satisfies the Fraunhofer condition and another workable set that satisfies the Fresnel condition, giving in both cases the diffractions patterns on A'-B' and the width of the central fringe.

Solution:
With both the two hypotheses the condition (ε is a small quantity less than one)

$$\frac{y-t}{b} < \varepsilon \qquad \tan\theta = \frac{y-t}{b} < \varepsilon \qquad \theta = \arctan(\frac{y-t}{b}) < \varepsilon$$

must be fulfilled (see Sec. 5.1.4).
When $t = 0$ we have the following maximum values (Fig. 5.47)

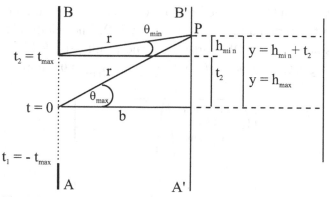

Fig. 5.47

$$\frac{y-t}{b} = \frac{h_{max}}{b} < \varepsilon \qquad \tan\theta_{max} = \frac{h_{max}}{b} \qquad \theta_{max} = \arctan(\frac{h_{max}}{b})$$

Assuming 0.01 as the greatest allowable value of ε, it follows for every value of t

$$h_{max} = 0.01b \qquad \theta_{max} = \arctan(0.01) = 0.57°$$

Hence, for example, we can assume

$$b_1 = 1000\,\text{mm} \qquad h_{1\,max} = 10\,\text{mm}$$

or

$$b_2 = 10000\,\text{mm} \quad h_{2\,\text{max}} = 100\,\text{mm}$$

With b_1 and $h_{1\text{max}}$ or with b_2 and $h_{2\text{max}}$ both the Fraunhofer and the Fresnel conditions are satisfied.

In the Fraunhofer situation the term t^2 is neglected in the formula

$$r = b + \frac{1}{2b}(y^2 - 2yt + t^2)$$

For example, using the smaller values $b = b_1$, $y = h_{1\text{max}}$ and $t_{\text{max}} = 0.1$ mm we have

$$y^2 - 2yt + t^2 \quad y^2 - 2yt = 10^2 - 2\cdot10\cdot0.1 = 80 \quad t^2 = 0.01$$

Hence we are in the Fraunhofer situation even if we use the Fresnel integral (see the formula (5.4) of the Sec. 5.1.4). The width of the central fringe is about 5mm. Of course the same result (Fig. 5.48) is expected using the greater values $b = b_2$, $y = h_{2\text{max}}$ and $t_{\text{max}} = 0.1$ mm

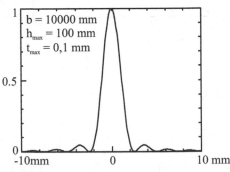

Fig. 5.48

$$y^2 - 2yt + t^2 \quad y^2 - 2yt = 10^4 - 2\cdot100\cdot0.1 = 9980 \quad t^2 = 0.01$$

The same plot is obtained using the square of the "sinc(α)" function (see Sec. 5.2.1) with

$$\alpha = \frac{\pi}{h}\sin\theta \quad h = 2t_{\text{max}} = 0.2\,\text{mm} \quad \theta = 0.57°$$

The abscissae in the plot of the square of the sinc(α) are, obviously, angles in the interval (-0.57,+0.57)°.

In the Fresnel situation using the smaller values $b = b_1$, $y = h_{1max}$ we must assume $t_{max} = 1$ mm or $t_{max} = 2$ mm if the term t^2 cannot be neglected

$$y^2 - 2yt + t^2 \qquad y^2 - 2yt = 10^2 - 2 \cdot 10 \cdot 1 = 80 \qquad t^2 = 1$$

$$y^2 - 2yt + t^2 \qquad y^2 - 2yt = 10^2 - 2 \cdot 10 \cdot 2 = 60 \qquad t^2 = 4$$

Using the formula (5.4) of the Sec. 5.1.4 the patterns of Fig. 5.49 are obtained and both the corresponding widths of the central fringe are about 8 mm.

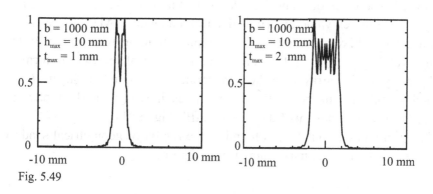

Fig. 5.49

5.2.20 Fresnel diffraction at an edge

On a vertical edge is incident normally (Fig. 5.50) a parallel beam of

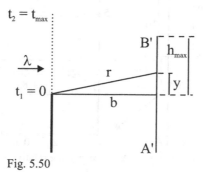

Fig. 5.50

monochromatic light ($\lambda = 0.5$ μ).

The screen A'-B' is 30 mm high ($y = h_{max} = \pm 15$ mm) and is distant $b = 4000$ mm from the edge. The value of t_{max} is assumed equal to 100 mm. Find the y positions for the minimum and the maximum of the relative intensity. Give the plot of the relative intensity varying y in the interval (-5, +10) mm. Observe the diffraction patterns for values of b smaller than 4000 mm and for very large value of b.

Solution:

The integral (see formula (5.4) in Sec. 5.1.4)

$$E(y) = D \int_{t_1}^{t_2} e^{i\pi \frac{(y-t)^2}{\lambda b}} \, dt$$

is used. The value of λ and b are known. With (Fig. 5.50) $t_1 = 0$ and $t_2 = t_{max}$ the complex value of $E(y)$ is determined using MATLAB numerical integration. Fixing N values of y, obtained dividing with a step Δy the interval ($-h_{max}, + h_{max}$), the corresponding array of the $E(y)$ is formed. Multiplying this array with the corresponding complex conjugate $E^*(y)$ and normalizing the result the relative intensities $I_r(y)$ can be plotted as function of the points y of the screen A'-B' (Fig. 5.51).

The minimum (about 1%) is found in the region of geometrical shadow at $y = -1.8$ mm. The maximum at $y = 1.2$ mm.

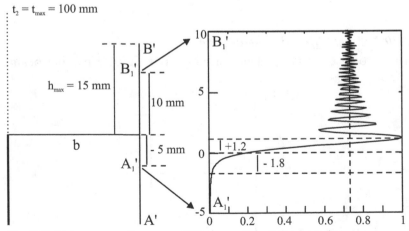

Fig. 5.51 Distances on the screen A'-B' versus the relative intensity

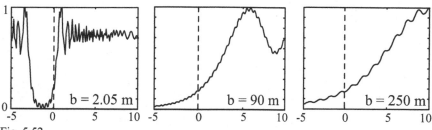

Fig. 5.52

Modified diffraction patterns are obtained (Fig. 5.52) varying b. With $b = 2050$ mm we have a maximum equal to one in the shadow region, too. With $b > 4000$ mm the maximum shifts to the right and with $b = 250$ meter the fringes disappear. With $b = 2000$ mm the relative intensity is equal to one everywhere on the screen from $y = -h_{max}$ to $y = h_{max} = 15$ mm.

Chapter 6

Photons and Moving Light Sources

6.1 Main Laws and Formulae

6.1.1 Introduction

A quantity we have particularly regarded and used in the previous chapter has been the intensity. This is the operational mode to consider the energy associated with the wavelike or corpuscular model of the light. If the energy is the preeminent quantity it is of secondary interest if to grasp the main feature of the light we need to think about it as a moving wave or as a moving entity called photon. The occurrence for a photon to have a linear and an angular momentum requires as a consequence that the mass is not necessary to own these properties. Certainly also in the corpuscular model the intensity is necessarily defined and used because what we can measure is always the energy striking, in a given time, some surface of a *detector*. The blackbody radiation analyzed by Planck gives rise to a distribution of the intensity per unit frequency, called spectral intensity. The problem of photoelectric effect resolved by Einstein concerns the intensity, too.

Another notable or distinctive feature of light is its insurmountable and invariable in the vacuum speed c. Consider, at astronomical or atomic distance, a source of radiation, stationary in the inertial system xyz, that moves with a relative speed v away from or toward another inertial system x'y'z'. If one observer A is stationary in the xyz and another B is stationary in x'y'z', the two observers can measure for the same radiation different frequencies, depending from the value and the sign of the ratio v/c, but they will always measures the same value for c, whichever the value and the sign of v/c. This follows from the Einstein postulate: *the speed of electromagnetic radiation in the vacuum is the same in all inertial systems*. We consider velocities of the light, of its source and of observers treating the Doppler effect.

6.1.2 The blackbody radiation

An enclosure of a body, a *blackbody*, when is heated at a uniform temperature T (in °K), contains radiation of different frequencies. The state of equilibrium for the spectral distribution of the energy, independent from the material that forms the cavity is a function of T only. Measurements show that the whole intensity is given by the Stefan-Boltzmann law

$$I = \sigma T^4 \qquad \sigma = 5.6742 \times 10^{-8} \frac{\text{watt}}{\text{m}^2 \, °\text{K}^4}$$

and that the maximum frequency is proportional to T (Wien's law)

$$\nu_{max} = AT$$

Using the spectral distribution of Planck the value of the constant A is

$$A = \frac{2.821 \, k}{h} = 5.88 \cdot 10^{10} \frac{\text{Hz}}{°\text{K}}$$

where $h = 6.626 \; 10^{-34}$ Joule sec and $k = 1.381 \; 10^{-23}$ Joule/ °K are the Planck and Boltzmann constants.

To provide a full account of these laws Planck introduced the quantum of the radiation, a *photon*, whose energy is proportional to the frequency

$$W_{ph} = h\nu$$

Using statistical methods Planck found for spectral distribution the following formula

$$I_\nu = \frac{2\pi h \nu^3}{c^2} \frac{1}{e^{\frac{h\nu}{kT}} - 1}$$

where *spectral intensity* I_v (see Appendix 1) has the unit of measure

$$\frac{\text{watt}}{\text{m}^2\,\text{Hz}} \quad \text{or} \quad \frac{\text{Joule}}{\text{m}^2}$$

with h and k defined just above and $c= 2.998\ 10^{\,8}$ m/sec is the speed of light in vacuum.

6.1.3 The photoelectric effect

If a beam of light or of radiation of higher frequencies (ultraviolet or X rays) hits a metal surface, electrons are emitted from the surface. Experimental results show that

a. there is a *minimum frequency* v_0 (called photoelectric threshold) allowing the emission of the electrons however is the intensity I of the radiation

b. *maximum energy* of the electrons is

$$W_{\max} = A(v - v_0)$$

where A will be assumed by Einstein equal to the Planck constant h. The maximum kinetic energy of the electron is function of the frequency v of the exciting radiation and of the frequency v_0 (the threshold characteristic of the material)

c. the *maximum velocity* of the electrons is independent from the intensity I of the radiation

d. the *number* of electrons emitted is proportional to the intensity I of the radiation

To provide a full account of these experimental facts Einstein assumed that in the photoelectric effect there is simply a collision between each photon and each electron and applied the principle of conservation of the energy to the collision. Hence it follows

$$hv = W_0 + eV_s$$

where on the left side we have the energy of the photon. On the right side we have first the energy W_0 (also called *work function*) used as work necessary to separate the electron from the metal surface. Then we have the quantity eV_s that is the maximum energy that an electron can acquire in the collision. The product eV_s is also called W_{max}. V_s is the potential difference required to reduce the velocity of the emitted electrons to zero. The definition of the intensity becomes

$$I = h\nu N \qquad N = \frac{P}{h\nu} = \frac{P\lambda}{hc}$$

where N is the number of photons per square meter and per second and P is power (watt) of the source emitting radiation.
The energy of a photon can also be written

$$W_{ph} = \frac{hc}{\lambda} \qquad hc = 1.24 \, \text{eV}^\circ$$

that can be used if, having the radiation wavelength λ in micron, the value of the energy in eV is requested.

6.1.4 Relativistic relations for linear momentum and energy of a particle and a photon

Defining m_0 the rest mass, for a particle moving with velocity v the mass m is

$$m = \gamma m_0 \qquad \gamma = \frac{1}{\sqrt{1 - \dfrac{v^2}{c^2}}}$$

For the electron $m_0 = 9.1 \ 10^{-31} \text{kg}$.
For a particle moving with velocity v, the linear momentum is

$$p = mv = \gamma m_0 v$$

and the energy

$$W = mc^2 = \gamma m_0 c^2$$

The kinetic energy of a particle moving with velocity v is

$$T = W - m_0 c^2 = mc^2 - m_0 c^2 = \gamma m_0 c^2 - m_0 c^2 = (\gamma - 1)m_0 c^2$$

where

$$m_0 c^2$$

is the rest energy of a particle of mass m_0.

Squaring the formula of the momentum we have

$$p^2 = \frac{m_0^2 \, v^2}{1 - \dfrac{v^2}{c^2}} \qquad p^2 (1 - \frac{v^2}{c^2}) = m_0^2 \, v^2 \qquad p^2 = (\frac{p^2}{c^2} + m_0^2) v^2$$

$$v^2 = \frac{p^2}{\dfrac{p^2}{c^2} + m_0^2}$$

Squaring the formula of the energy we have

$$W^2 = \frac{m_0^2 \, c^4}{1 - \dfrac{v^2}{c^2}} \qquad W^2 (1 - \frac{v^2}{c^2}) = m_0^2 \, c^4 \qquad W^2 - m_0^2 \, c^4 = E^2 \, \frac{v^2}{c^2}$$

$$v^2 = \frac{c^2 \, (W^2 - m_0^2 \, c^4)}{W^2}$$

Equating the two formulae of the velocity v, after some algebraic operations

$$\frac{c^2 \, (E^2 - m_0^2 \, c^4)}{E^2} = \frac{p^2}{\dfrac{p^2}{c^2} + m_0^2} \qquad \frac{(E^2 - m_0^2 \, c^4)}{E^2} = \frac{p^2}{p^2 + c^2 m_0^2}$$

$$(E^2 - m_0^2 \, c^4)(p^2 + c^2 m_0^2) = E^2 \, p^2$$

$$E^2 \, p^2 + E^2 \, c^2 m_0^2 - m_0^2 \, c^4 p^2 - m_0^2 \, c^4 c^2 m_0^2 = E^2 \, p^2$$

$$E^2 \, c^2 m_0^2 - m_0^2 \, c^4 p^2 - m_0^2 \, c^4 c^2 m_0^2 = 0$$

we have

$$W^2 = c^2 p^2 + m_0 c^4 \qquad (6.1)$$

that can be rewritten, remembering the formula for kinetic energy, as

$$c^2 p^2 = W^2 - m_0^2 c^4 = (W + m_0 c^2)(W - m_0 c^2) \qquad (6.2)$$

$$c^2 p^2 = (W - m_0 c^2 + 2m_0 c^2)(W - m_0 c^2) = (T + 2m_0 c^2)T \qquad (6.3)$$

For the photon of the frequency v and energy

$$W_{ph} = hv$$

the linear momentum follows from (6.1)

$$W^2 = c^2 p^2 \qquad hv = cp \qquad p = \frac{hv}{c} \qquad (6.4)$$

because for a photon the rest mass is zero.

6.1.5 The Compton's effect

We assume a frame of reference xy where the electron is at rest in the origin (Fig. 6.1). Moving along the x direction, a photon of frequency v is scattered by an electron of rest mass m_0. After collision the photon has frequency v' and the electron has a speed v. Their directions of motions subtend angles θ and φ with the x axis. The kinematics of the collision is determined by the conservation principles of energies and of the components of the momentum along the x and the y axes. Hence we have

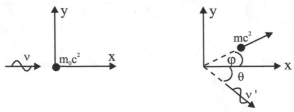

Fig. 6.1

$$h\nu + m_0 c^2 = h\nu' + W_e \qquad \rightarrow \qquad h(\nu - \nu') = W_e - m_0 c^2 \qquad (6.5)$$

$$\frac{h\nu}{c} = \frac{h\nu'}{c}\cos\theta + p_e \cos\varphi \qquad 0 = \frac{h\nu'}{c}\sin\theta - p_e \sin\varphi$$

where W_e and p_e are the energy and the momentum of the electron. Rearranging the previous formulae of the components of the momentum

$$c\,p_e \cos\varphi = h(\nu - \nu'\cos\theta) \qquad c\,p_e \sin\varphi = h\nu'\sin\theta$$

we have

$$c^2 p_e^2 = h^2 (\nu^2 - 2\nu\nu'\cos\theta + \nu'^2) \qquad (6.6)$$

From (6.2) and (6.5) we have

$$c^2 p_e^2 = (W_e + m_0 c^2)(W_e - m_0 c^2) = (h(\nu - \nu') + 2m_0 c^2)\, h(\nu - \nu') \quad (6.7)$$

Equating the right-hand sides of the last two formulae and after some simple algebraic operations, we obtain the Compton formula linking the frequencies or the wavelengths of the photon, before and after the collision

$$\frac{\nu - \nu'}{\nu\nu'} = \frac{h}{m_0 c^2}(1 - \cos\theta) \qquad \lambda' - \lambda = \frac{h}{m_0 c}(1 - \cos\theta) \qquad (6.8)$$

with

$$\frac{h}{m_0 c} = 2.4263 \text{ pm} \qquad 1 \text{ pm} = 10^{-12}\text{m}$$

6.1.6 Light Doppler effect

6.1.6a The light Doppler effect: classical approach
We assume that a source moves with the velocity v and emits light that has frequency v and velocity c. The observer, moving with velocity v_0,

measures a frequency v' for the light when this reaches him. Directions are linear, along the direction A-B, and magnitudes are constant for all the three velocities (Fig. 6.2).

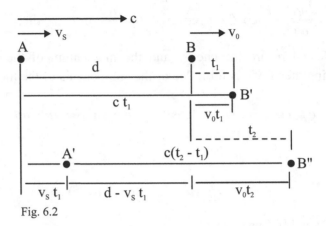

Fig. 6.2

At time $t = 0$ the source S, radiating light, moves from A and the observer begins to move from B. At the time t_1 the observer is in B' when he is reached from the light emitted from S at time $t = 0$. Hence

$$ct_1 = d + v_0 t_1 \qquad d = (c - v_0)t_1 \tag{6.9}$$

At time t_2 the observer is in B" when he is reached from the light emitted from S when it was in A'. Hence, using (6.9) for d,

$$c(t_2 - t_1) = d - v_S t_1 + v_0 t_2 \qquad c(t_2 - t_1) = ct_1 - v_0 t_1 - v_S t_1 + v_0 t_2$$

or

$$c(t_2 - t_1) = (c - v_S)t_1 + v_0(t_2 - t_1)$$

From the last formula it follows

$$\frac{t_1}{t_2 - t_1} = \frac{c - v_0}{c - v_S} \tag{6.10}$$

The observer perceives through the time $t_2 - t_1$ a number N of waves of a different frequency ν' equal to the N waves emitted, with frequency ν, from the source through the time t_1. Hence, with *periods T and T'* as the basic time for counting the waves,

$$\frac{t_1}{T} = \frac{t_2 - t_1}{T'} \qquad t_1 \nu = (t_2 - t_1)\nu'$$

Hence from the previous formula and (6.10) we have

$$\nu' = \frac{t_1}{t_2 - t_1} = \frac{(c - v_0)}{(c - v_S)}\nu \qquad (6.11)$$

Formula (6.11) can also be written

$$\nu' = \frac{(1 - \frac{v_0}{c})}{(1 - \frac{v_S}{c})}\nu = \nu(1 - \frac{v_0}{c})(1 - \frac{v_S}{c})^{-1} = \nu(1 + \frac{v_S}{c} - \frac{v_0}{c}) \qquad (6.12)$$

using only the first two terms of the power series for denominator and omitting the term $v_S v_0/c^2$.

In (6.11) and (6.12) v_0 and v_S can have equal or different sign leading to situations of more complexity. The simplest cases occurs when either v_0 or v_S is zero.

If the *source is stationary* in A ($v_S = 0$) and the observer is moving to the right of B ($v_0 > 0$) from (6.11) or (6.12) we have

$$\nu' = (1 - \frac{v_0}{c})\nu \qquad (6.13)$$

If the *source is stationary* in A ($v_S = 0$) and the observer is moving to the left of B ($v_0 < 0$) from (6.11) or (6.12) we have

$$\nu' = (1 + \frac{v_0}{c})\nu \qquad (6.14)$$

If the *observer is stationary* in B ($v_0 = 0$) but the source of the light is moving from A to B ($v_S > 0$) from (6.11) or (6.12) we have

$$\nu' = (1 + \frac{v_S}{c})\nu \qquad (6.15)$$

If the *observer is stationary* in B ($v_0 = 0$) but the source of the light is moving to the left of A ($v_S < 0$) from (6.11) or (6.12) we have

$$v' = (1 - \frac{v_S}{c})v \qquad (6.16)$$

6.1.6b The light Doppler effect: relativistic approach

From the two basic postulates of the special theory of relativity the following Lorentz transformations follow

$$x = \gamma(x' + vt') \quad y = y' \quad z = z' \quad t = \gamma(t' + \frac{v}{c^2}x')$$

$$x' = \gamma(x - vt') \quad y = y' \quad z = z' \quad t' = \gamma(t' - \frac{v}{c^2}x') \quad \gamma = \frac{1}{\sqrt{1 - \frac{v^2}{c^2}}} \qquad (6.17)$$

One inertial system has coordinate xyz and another has coordinate $x'y'z'$ (Fig. 6.3). The inertial systems are moving with relative constant speed v. The direction of axes x and x' and of v is the same. The sign of v must be

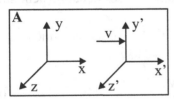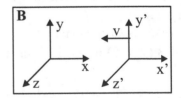

Fig. 6.3

changed in (6.17) if the condition of box **B** of Fig. 6.3 is verified. According to the first postulate the electromagnetic laws are effective in all inertial systems. The second postulates states that the speed of light in the vacuum is the same in all inertial systems. We assume the light represented as a harmonic wave of unitary magnitude. A light source present in the inertial system xyz, where it emits a beam of frequency v that subtends an angle θ with the x axis, is seen by an observer in the inertial system $x'y'z'$ to have the frequency v' and subtend the angle θ' with x' axis (Fig. 6.4)

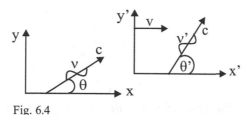

Fig. 6.4

$$f'=\sin(\frac{2\pi}{\lambda'}r'-\omega't')=\sin(\frac{2\pi\nu'}{c}r'-2\pi\nu't')=\sin[2\pi\nu'(\frac{r'}{c}-t')]=$$

$$=\sin[2\pi\nu'(\frac{x'\cos\theta'+y'sen\theta'}{c}-t')]=$$

$$=\sin[2\pi(\frac{\nu'\cos\theta'}{c}x'+\frac{\nu'\sin\theta'}{c}y'-\nu't')] \qquad (6.18)$$

In the *xyz* inertial system it will be

$$f=\sin[2\pi\nu(\frac{x\cos\theta+y sen\theta}{c}-t)] \qquad (6.19)$$

Using the Lorentz transformations (6.17) in the argument of the sine function of formula (6.19) we have

$$\nu(\frac{x\cos\theta+y\sin\theta}{c}-t)=\nu[\frac{\gamma(x'+vt')\cos\theta+y'\sin\theta}{c}-\gamma(t'+\frac{v}{c^2}x')]=$$

$$=\nu[\frac{\gamma(x'+vt')\cos\theta}{c}+\frac{y'\sin\theta}{c}-\gamma(t'+\frac{v}{c^2}x')]=$$

$$=\nu[\frac{\gamma}{c}(\cos\theta-\frac{v}{c})x'+\frac{\sin\theta}{c}y'-\gamma(1-\frac{v}{c}\cos\theta)t'] \qquad (6.20)$$

Equating the coefficients of *x'*, *y'* and *t'* of (6.18) with those in (6.20) we obtain

$$\nu'\cos\theta'=\nu\gamma(\cos\theta-\frac{v}{c}) \qquad \nu'\sin\theta'=\nu\sin\theta \qquad \nu'=\nu\gamma(1-\frac{v}{c}\cos\theta) \quad (6.21)$$

Dividing the first by the last of (6.21) the relation between θ and θ' is obtained

$$\cos\theta'=\frac{\cos\theta-\dfrac{v}{c}}{1-\dfrac{v}{c}\cos\theta} \qquad (6.22)$$

First we consider the case of *longitudinal Doppler effect* when $\theta = 0°$ (Fig. 6.5). From (6.22) and the last of (6.21) it follows

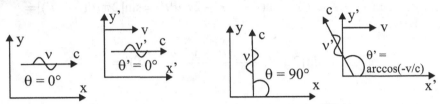

Fig. 6.5 Longitudinal ($\theta = 0°$) and transverse ($\theta = 90°$) Doppler Effect

$$\cos\theta'=1 \quad \theta'=\theta=0 \quad \nu'=\nu\gamma(1-\frac{v}{c})=\nu\frac{1-\dfrac{v}{c}}{\sqrt{1-\dfrac{v^2}{c^2}}} \qquad (6.23)$$

If the square of v/c can be neglected the previous formula for ν' becomes

$$\nu'=\nu(1-\frac{v}{c}) \qquad (6.24)$$

that includes the four formulae (6.13), (6.14), (6.15) and (6.16) of the classical approach.

With $\theta = 90°$ or $\theta = -90°$ we have the case of *transverse Doppler effect* (Fig. 6.5). Then we have

$$\theta=\pm90° \quad \cos\theta'=-\frac{v}{c} \quad \theta'=\arccos(-\frac{v}{c}) \quad \nu'=\nu\gamma=\frac{\nu}{\sqrt{1-\dfrac{v^2}{c^2}}} \qquad (6.25)$$

that can also be written, using the first two terms of the power series expansion

$$\nu'=\nu[1+\frac{1}{2}\frac{v^2}{c^2}+\frac{3}{8}(\frac{v^2}{c^2})^2+...]=\nu(1+\frac{1}{2}\frac{v^2}{c^2}) \qquad (6.26)$$

Now the value of v', depending from the square of v/c, is more difficult to observe and measure.

The inverse formulae to calculate the values of v and θ assigned v' and θ' can be obtained. In fact from (6.22) we have

$$\cos\theta = \frac{\dfrac{v}{c}+\cos\theta'}{1+\dfrac{v}{c}\cos\theta'} \qquad (6.27)$$

Using this value of $\cos\theta$ from the last formula in (6.21) we obtain

$$\nu = \nu'\gamma(1+\frac{v}{c}\cos\theta') \qquad (6.28)$$

If the inertial systems x'y'z' and xyz are moving toward each other (box **B** of Fig. 6.3) *the sign of* v *must be changed in all the previous formulae.*

6.1.6c Classical approach as a case of relativistic longitudinal Doppler effect

The relativistic longitudinal Doppler effect contains as a special case the classical approach of the Doppler effect. We rewrite the formula (6.24)

$$\nu' = \nu(1-\frac{v}{c}) \qquad (6.24)$$

where v is the relative velocity between the two inertial systems and replaces the velocities v_S and v_0 of the classical model (Fig. 6.6).
For a *positive* value of v (the two inertial systems are moving away from one another) formula (6.24) is equivalent to (6.13) and (6.16).
For a *negative* value of v (the two inertial systems are moving toward each other) formula (6.24) is equivalent to (6.14) and (6.15).

The two systems are receding   The two systems are approaching

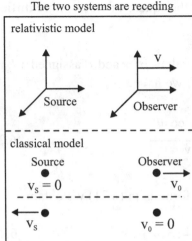

Fig. 6.6

6.2 Problems

6.2.1 Planck 1

Using the spectral intensity of a blackbody radiation I_ν (Fig. 6.7), find for temperatures $T_1 = 300$, $T_2 = 1000$, $T_3 = 2400$, $T_4 = 4500$ and $T_5 = 6000°K$

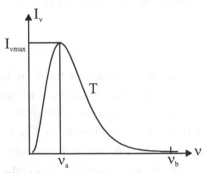

Fig. 6.7

the maximum value of the frequency $\nu_{b\,max}$ that allows to observe an appreciable value of I_ν different from zero. Where is placed this value in the electromagnetic spectrum? Determine the values of

$$g_i = \frac{h\nu_{ai}}{kT_i} \qquad i = 1, 2, 3, 4, 5$$

where $v_{a\,i}$ is the frequency, when temperature is T_i, for which I_v has the maximum value $I_{v\,max}$.

Define how the spectral intensities are displaced among the infrared, visible and ultraviolet regions and the ratio of the intensity of the visible radiation to the intensity of total electromagnetic radiation for the assigned temperatures. Assume as limit frequency for the visible region $v_1 = 4.3 \cdot 10^{14}$ Hz for the red and $v_2 = 7.4 \cdot 10^{14}$ Hz for the violet light.

Solution:

Plotting the Planck's law

$$I_v = \frac{2\pi h v^3}{c^2} \frac{1}{e^{\frac{hv}{kT}} - 1}$$

for the lower, $T_1 = 300°$, and the higher value of temperature, $T_5 = 6000°K$ (Fig. 6.8), we find

$$v_{b\,max} = 1.5 \cdot 10^{15}\,\text{Hz}$$

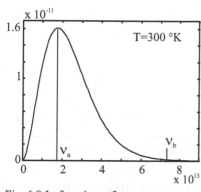
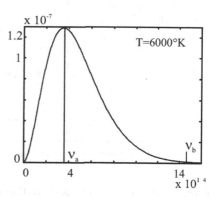

Fig. 6.8 I_v function of v

This is an ultraviolet frequency greater than $7.4 \cdot 10^{14}$ Hz (the limit of the visible violet light).

Using the $v_{a\,i}$ corresponding to the assigned temperatures (Table 6.1) we have a constant value for g,

$$g_i = \frac{h\,v_{ai}}{k\,T_i} = 4.798 \cdot 10^{-11} \frac{v_{ai}}{T_i} = 2.82 \qquad i = 1,2,3,4,5$$

Table 6.1

T (°K)	300	1000	2400	4500	6000
v_a (x10^{13})	1.76	5.88	14.1	26.5	35.3
v_a/T (x10^{10})	5.87	5.88	5.88	5.88	5.88
g		2.81	2.82	2.82	2.82	2.82

Hence we may write the linear relation (Wien's law)

$$\nu_a = \frac{gk}{h}T = 5.88 \cdot 10^{10}\,T$$

The allocation of the spectrum among the infrared, visible and ultraviolet regions varying temperatures is given in Fig. 6.9. The ratio of the intensity in the visible region to the whole intensity, for an assigned tempera-

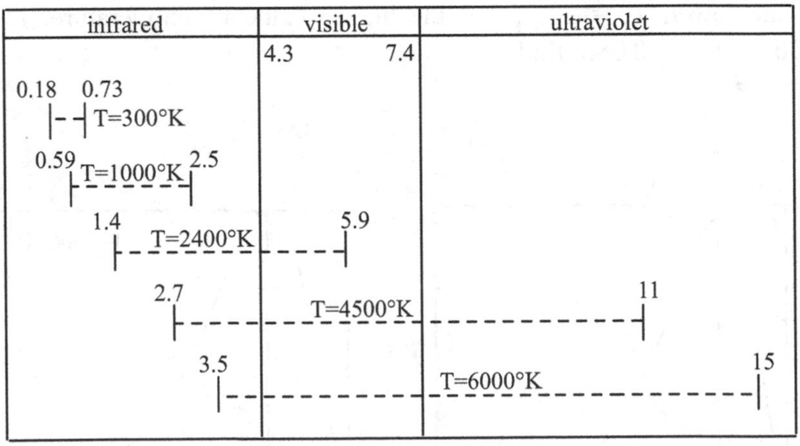

Fig. 6.9 Numbers without unit of measure are frequencies (x10^{14} Hz)

ture, is given with the aid of the numerical integration of the spectral intensity

$$I(T) = \int_{\nu_1}^{\nu_2} \frac{2\pi h\nu^3}{c^2} \frac{1}{e^{\frac{h\nu}{kT}} - 1}\,d\nu$$

with $\nu_1 = 4.3 \cdot 10^{14}$ Hz and $\nu_2 = 7.4 \cdot 10^{14}$ Hz when the intensity in the visible region is calculated and $\nu_1 = 10$ Hz and $\nu_2 = 10^{16}$ Hz when the intensity due to the whole spectrum, for an assigned temperature, is calculated (Fig. 6.10).

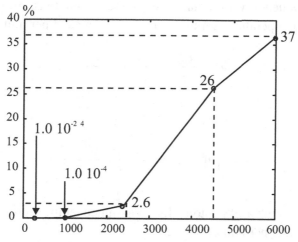

Fig. 6.10 % of visible to the whole intensity versus T (°K)

6.2.2 Planck 2

Using the spectral intensity I_ν of a blackbody radiation, find for temperatures $T_1 = 300$, $T_2 = 1000$, $T_3 = 2400$, $T_4 = 4500$ and $T_5 = 6000°$K the maximum number of photon, per unit of time and per unit of surface, radiated from the blackbody assuming that the radiation is in the frequency range between $\nu_{min} = 10^6$ Hz and $\nu_{max} = 10^{16}$ Hz (Fig. 6.11).

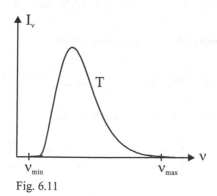

Fig. 6.11

Solution:
Dividing the spectral intensity I_ν by the energy of a photon, corresponding to a frequency ν we obtain the number of photons, of a given frequency and temperature, emitted per unit of time and per unit of surface

$$\Phi_\nu = \frac{I_\nu}{h\nu} = \frac{2\pi\nu^2}{c^2}\frac{1}{e^{\frac{h\nu}{kT}}-1}$$

The integral

$$\Phi = \int \Phi_\nu \, d\nu = \frac{2\pi}{c^2}\int \frac{\nu^2}{e^{\frac{h\nu}{kT}}-1}\, d\nu$$

will give the number of photons, for an assigned temperature, emitted per
unit of time and per unit of surface
Putting

$$\frac{h\nu}{kT}=x \qquad \nu = \frac{kT}{h}x \qquad d\nu = \frac{kT}{h}dx$$

we have

$$\Phi_T = \frac{2\pi}{c^2}\left(\frac{kT}{h}\right)^3 \int \frac{x^2}{e^x-1}\, dx$$

where the subscript emphasizes that it is dependent from the temperature
T.
For the assigned limits of the frequency range

$$\nu_{min} = 1.0\cdot10^6\,\text{Hz} \qquad\qquad \nu_{max} = 1.0\cdot10^{16}\,\text{Hz}$$

the corresponding values of the variable x are

$$x = \frac{h\nu}{kT} \qquad x_{min} = \frac{h\,\nu_{min}}{k\,T} \qquad x_{max} = \frac{h\,\nu_{max}}{k\,T} \qquad \frac{h}{k} = 4.8\times10^{-11}$$

Hence we have

$$\Phi_T = \frac{2\pi}{c^2}\left(\frac{kT}{h}\right)^3 \int_{x_{min}}^{x_{max}} \frac{x^2}{e^x-1}\, dx$$

Assuming $x_{min} = 0$ for all the five assigned temperatures whereas the corresponding x_{max} are 1559, 480, 200, 107 and 80 for the corresponding values of temperatures the integral has always the same value

$$\int_{x_{min}}^{x_{max}} \frac{x^2}{e^x - 1} dx = 2.40$$

Hence we have

$$\Phi_T = \frac{2\pi}{c^2} \left(\frac{kT}{h}\right)^3 \cdot 2.4 = \frac{4.8\pi k^3}{c^2 h^3} T^3 = 1.52 \cdot 10^{15} T^3$$

The whole number of photons (but of different energy) emitted by the blackbody per unit of time and per unit of surface, proportional to third power of T, are (multiplied by 10^{22})

$$\Phi_1 = 4.1 \quad \Phi_2 = 150 \quad \Phi_3 = 2100 \quad \Phi_4 = 14000 \quad \Phi_5 = 33000$$

6.2.3 Planck 3

Consider the Sun as a blackbody spherical radiator. It has a radius $r = 6.96$ 10^8m, a temperature $T = 5800$ °K and is distant $R = 1.496 \ 10^{11}$m from the Earth (Fig. 6.12). We suppose R *constant*. Find the intensity I_S of the solar

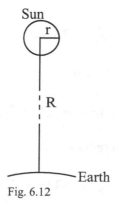

Sun

Fig. 6.12

radiation, the partial intensity I_V due to the visible domain and the corresponding values of the total and partial powers W_S and W_V. Assuming there isn't absorption of radiation by the gaseous substances surrounding the Earth find the solar intensity I_E on the Earth and its partial value I_{EV} due to the visible radiation.

Solution:
The intensity of the radiation emitted from the Sun is given by the Stefan-Boltzmann law

$$I_S = \sigma T^4 = 6.4212 \cdot 10^7 \text{ watt/m}^2 \qquad \sigma = 5.6742 \cdot 10^{-8} \frac{\text{watt}}{\text{m}^2 \, {}^\circ \text{K}^4}$$

or may be numerically calculated by the integral

$$I_S = \int_{\nu 1}^{\nu 2} I_\nu \, d\nu = 6.4134 \cdot 10^7 \text{ watt/m}^2$$

assuming $T = 5800^\circ\text{K}$, $\nu_1 = 10^3$ Hz and $\nu_2 = 1.5 \cdot 10^{15}$ Hz (see the first question of Sec. 6.2.1).
The intensity of the visible radiation emitted from the Sun is

$$I_V = \int_{\nu 1}^{\nu 2} I_\nu \, d\nu = 2.3050 \cdot 10^7 \text{ watt/m}^2$$

assuming $T = 5800^\circ\text{K}$, $\nu_1 = 4.3 \cdot 10^{14}$ Hz and $\nu_2 = 7.4 \cdot 10^{14}$ Hz (see Sec. 6.2.1).
Their ratio is

$$\frac{I_V}{I_S} = 35.94\%$$

The spherical surface of the Sun is

$$S_S = 4\pi r^2 = 6.1 \cdot 10^{18} \text{m}^2$$

Hence the power of the Sun radiation is

$$W_S = I_S \cdot S_S = 3.9 \cdot 10^{26} \text{watt}$$

The partial power due to the visible radiation of the Sun is

$$W_V = I_V \cdot S_S = 1.4 \cdot 10^{26} \text{watt}$$

The great spherical surface (Fig. 6.13) whose radius is the distance R from the Sun to the Earth has the area

$$S_E = 4\pi R^2 = 2.8 \cdot 10^{23} \text{m}^2$$

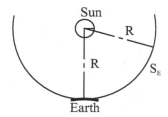

Fig. 6.13

The intensity of the Sun radiation on every piece of the surface S_E and on the Earth (a small part of the area S_E) is

$$I_E = \frac{W_S}{S_E} = 1390 \text{ watt/m}^2$$

This is usually called *solar constant*.
The partial intensity due to the visible radiation irradiated from the Sun on Earth is

$$I_{EV} = \frac{W_V}{S_E} = 498 \text{ watt/m}^2$$

6.2.4 Photons 1
A point source O radiates monochromatic light (λ = 589 nm) with a power P = 100 watt. Find the radius R_1 of the spherical surface where there is the transit of one photon per cm^2 and the radius R_2 of the spherical surface where there is a density δ of one photon per sec and per cm^3.

Solution:
If N is the number of photons emitted per unit of time (ph/sec) and $h\nu$ is the energy of a photon we can write

$$P = h\nu N \qquad N = \frac{P}{h\nu} = \frac{P\lambda}{hc} = 2.97 \cdot 10^{20} \text{ph/sec}$$

The number N of photons must reach the surface (Fig. 6.14) and its portion Su = 1 cm² must be reached by Nu =1 ph/sec.

$$S = 4\pi R_1^2$$

Fig. 6.14

Hence

$$\frac{N}{S} = \frac{N_u}{S_u} = \frac{1\text{ph/sec}}{1\text{cm}^2} = \frac{1\text{ph/sec}}{10^{-4}\text{m}^2} \qquad N = 4\pi R_1^2 \frac{\text{ph/sec}}{10^{-4}\text{m}^2}$$

$$R_1 = \sqrt{\frac{N 10^{-4}\text{m}^2}{4\pi\text{ph/sec}}} = \sqrt{\frac{2.97 \cdot 10^{20} 10^{-4}\text{m}^2}{4\pi}} = 4.86 \cdot 10^7\,\text{m}$$

The density is

$$\delta = \frac{N}{V} = \frac{N}{Sb} = \frac{N}{Sct} = \frac{N}{Sc \cdot 1\text{sec}} = \frac{N}{4\pi R_2^2 c \cdot 1\text{sec}}$$

where we have put $t = 1$ sec and assumed $S_1 = S_2$ (Fig. 6.15). This density must be equal to

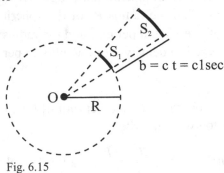

Fig. 6.15

$$\delta = 1 \frac{\text{ph/sec}}{\text{cm}^3}$$

Hence, from the previous two formulae and remembering the value of N, it follows

$$\frac{2.97 \cdot 10^{20}}{4\pi R_2^{\,2}\, 2.988 \cdot 10^8\,\text{m}} = \frac{1}{10^{-6}\,\text{m}^3}$$

or

$$R_2^{\,2} = \frac{2.97 \cdot 10^6}{4\pi\, 2.998}\,\text{m}^2 \qquad R_2 = \sqrt{\frac{2.97}{4\pi\, 2.998}}\, 10^3\,\text{m} = 280\,\text{m}$$

6.2.5 Photons 2

Two point sources, of power $P = 400$ watt, have respectively $\lambda_V = 400$ nm and $\lambda_R = 400$ nm. Find the difference $N_R - N_V$ between the numbers of photons per second emitted from the two sources. Find the intensities at a distance $R = 2$ m as number of photons per sec and per m^2 and as watt/m^2.

Solution:

If N is the number of photons per second and $h\nu$ is the energy of every photon we may write the power as

$$P = N h\nu = N h \frac{c}{\lambda}$$

hence

$$N = \frac{P\lambda}{hc}$$

We find the value of the product hc

$$hc = 6.626 \times 10^{-34}\,\text{Joule} \cdot \text{sec} \cdot 2.998 \cdot 10^8\,\frac{\text{m}}{\text{sec}} = 1.99 \cdot 10^{-25}\,\text{Joule} \cdot \text{m}$$

Therefore the values of N for the two wavelengths are

$$N_v = \frac{P\lambda_V}{hc} = \frac{400\,\text{Joule/sec} \cdot 400 \cdot 10^{-9}\,\text{m}}{1.99 \cdot 10^{-25}\,\text{Joule} \cdot \text{m}} = 8.1 \cdot 10^{20}\,\text{ph/sec}$$

and

$$N_R = \frac{P\lambda_R}{hc} = \frac{2.8 \cdot 10^{-4}}{1.99 \cdot 10^{-25}} = 1.4 \cdot 10^{21} \, \text{ph/sec}$$

Their difference is

$$diff = N_R - N_V = 6.7 \cdot 10^{20} \, \text{ph/sec}$$

The intensities are

$$I_V = \frac{N_V}{S} = \frac{8.1 \cdot 10^{20} \, \text{ph/sec}}{4\pi R^2} = \frac{8.1 \cdot 10^{20} \, \text{ph/sec}}{50.27 \, \text{m}^2} = 1.6 \cdot 10^{19} \frac{\text{ph/sec}}{\text{m}^2}$$

$$I_R = \frac{N_R}{S} = \frac{1.4 \cdot 10^{21} \, \text{ph/sec}}{50.27 \, \text{m}^2} = 2.8 \cdot 10^{19} \frac{\text{ph/sec}}{\text{m}^2}$$

Their values, in standard unit of measure, are, as expected, equal

$$I_V = 1.6 \cdot 10^{19} \frac{hc/\lambda_V}{\text{sec} \cdot \text{m}^2} = 1.6 \cdot 10^{19} \frac{1.99 \cdot 10^{-25} \, \text{Joule} \cdot \text{m}}{400 \cdot 10^{-9} \text{msec} \cdot \text{m}^2} = 7.96 \, \text{watt/m}^2$$

$$I_R = 2.8 \cdot 10^{19} \frac{hc/\lambda_R}{\text{sec} \cdot \text{m}^2} = 2.8 \cdot 10^{19} \frac{1.99 \cdot 10^{-25} \, \text{Joule} \cdot \text{m}}{700 \cdot 10^{-9} \text{sec} \cdot \text{m}^2} = 7.96 \, \text{watt/m}^2$$

6.2.6 Photons 3
A beam of monochromatic light ($\lambda = 200$ nm) is incident on a metal surface whose work function is $W_0 = 4.2$ eV. Find the energy W_{max} of the electrons ejected from the metal surface, the stopping potential V_s and the maximum wavelength λ_0 corresponding to the photoelectric threshold.

Solution:
From the basic formula of photoelectric effect

$$h\nu = W_0 + eV_s$$

we have first the stopping potential

$$V_s = \frac{hc}{\lambda e} - \frac{W_0}{e} = \frac{1}{e}(\frac{1.2415}{0.2\mu} \text{eV} \cdot ^\circ - 4.2 \text{ eV}) = (\frac{1.2415}{0.2} - 4.2) \text{V} = 2.0 \text{V}$$

and then the energy W_{max}

$$W_{max} = eV_s = 2.0\,eV$$

There isn't photoelectric effect if

$$h\nu_0 < W_0 \qquad \rightarrow \qquad \frac{hc}{\lambda_0} < W_0$$

Hence the wavelength for photoelectric effect threshold is

$$\lambda_0 > \frac{hc}{W_0} = \frac{1.2415\ eV \cdot {}^{\circ}}{4.2\,eV} = 0.296^{\circ}$$

6.2.7 Photons 4

A sodium surface has a stopping potential $V_{s\,1} = 1.85$ Volt when a monochromatic ($\lambda_1 = 300$ nm) beam is incident on it. If a different monochromatic ($\lambda_2 = 400$ nm) beam is incident on the surface the stopping potential is $V_{s\,2} = 0.82$ Volt. Find the value of the work function W_0 without using the constant h and c; then, using the value found for W_0 and the constant c, determine the value of the Planck's constant h.

Solution:
Applying the basic formula two times, we have

$$\frac{hc}{\lambda_1} = W_0 + eV_{s1} \qquad \frac{hc}{\lambda_2} = W_0 + eV_{s2}$$

or

$$hc = \lambda_1(W_0 + eV_{s1}) \qquad hc = \lambda_2(W_0 + eV_{s2})$$

Equating the right-hand terms of the two previous formulae we have

$$W_0 = e\frac{\lambda_1 V_{s1} - \lambda_2 V_{s2}}{\lambda_2 - \lambda_1} = 2.27\,eV$$

The value of h can be retrieved using the first or the second application of the basic formula. For example, using the first we have

$$\frac{hc}{\lambda_1}=W_0+eV_{s1} \qquad h=\frac{\lambda_1}{c}(W_0+eV_{s1})=4.12\cdot10^{-15}\,\text{eV}\cdot\text{sec}$$

More accurate values for h and W_0 are, respectively, 4.14 10^{-15} eVsec and 2.28 eV.

6.2.8 Compton 1

A beam of photons of wavelength $\lambda = 12$ pm collides with electrons of rest mass m_0. After collision the photons, scattered at an angle $\theta = 52°$ from its initial direction, have a wavelength λ' (Fig. 6.16). Find λ', the energy lost by the photons and gained by the electrons, and the scattering direction φ of the electrons.

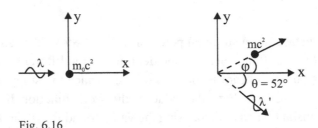

Fig. 6.16

Solution:
With

$$A=\frac{h}{m_0 c}=2.4263 \text{ pm}$$

the new wavelength is

$$\lambda'=\lambda+A(1-\cos\theta)=12.9 \text{ pm}$$

Applying the principle of conservation of energy we have

$$W_{ph}+m_0 c^2 =W_{ph}'+W_e$$

$$W_e - m_0 c^2 = W_{ph}-W_{ph}'=hc\frac{\lambda'-\lambda}{\lambda'\lambda}=7.46 \text{ keV}$$

Applying the principle of conservation of linear momentum

$$\frac{h\nu}{c}=\frac{h\nu'}{c}\cos\theta+p\cos\varphi \qquad 0=\frac{h\nu'}{c}\sin\theta-p\sin\varphi$$

we can find the angle φ

$$p\cos\varphi=\frac{h\nu}{c}-\frac{h\nu'}{c}\cos\theta=\frac{h}{c}(\nu-\nu'\cos\theta) \qquad p\sin\varphi=\frac{h\nu'}{c}\sin\theta$$

$$\tan\varphi=\frac{\nu'\sin\theta}{\nu-\nu'\cos\theta} \qquad \varphi=\arctan(\frac{\nu'\sin\theta}{\nu-\nu'\cos\theta})=59.6°$$

6.2.9 Compton 2
A beam of photons of frequency $\nu=2.5\ 10^{19}$ Hz is scattered by electrons of rest mass m_0. After collision the photons are scattered at 90° from the initial direction with an energy loss equal to 0.034 of the rest energy $m_0 c^2$ of the electrons. Find its frequency ν' after the collision and the decrease $\Delta\nu_{ph}$ in eV of energy of the photon. Calculate the value of the velocity and the angle φ of the direction of the electrons after the collision (Fig. 6.17).

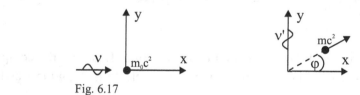

Fig. 6.17

Solution:
From the Compton formula

$$\frac{\nu-\nu'}{\nu\nu'}=\frac{h}{m_0 c^2}(1-\cos\theta) \qquad \nu'=\frac{\nu}{1+\nu A(1-\cos\theta)} \qquad A=\frac{h}{m_0 c^2}$$

for $\theta = 90°$ we have

$$\nu'=\frac{\nu}{1+\nu A}=2.1\cdot10^{19}\,\text{Hz}$$

The energy loss by the photon is

$$\Delta W_{ph} = h(\nu - \nu') = 1.74 \cdot 10^4 \text{eV}$$

Before the collision we have

$$W_{ph} = h\nu \qquad P_{phx} = \frac{h\nu}{c} \qquad .P_{phy} = 0$$

for the energy and the momentum of the photon and

$$W_e = m_0 c^2 \qquad p_{ex} = p_{ey} = 0$$

for the energy and the momentum of the electron. After collision we have the

$$W'_{ph} = h\nu' \qquad p'_{phx} = 0 \qquad p'_{phy} = \frac{h\nu'}{c}$$

for the energy and the momentum of the photon and

$$W_e' = \gamma m_0 c^2 \qquad \gamma = \frac{1}{\sqrt{1 - \dfrac{v^2}{c^2}}} \qquad p'_{ex} = p'_e \cos\varphi \qquad p'_{ey} = p'_e \sin\varphi$$

for the energy and the momentum of the electron. Equating the components of the momenta of the photon and of the electron along the x and y axes we have

$$\frac{h\nu}{c} = p'_e \cos\varphi \qquad 0 = \frac{h\nu'}{c} - p'_e \sin\varphi \qquad \rightarrow \qquad \frac{h\nu'}{c} = p'_e \sin\varphi \qquad (6.29)$$

and

$$\tan\varphi = \frac{\nu'}{\nu} \qquad \varphi = \arctan(\frac{\nu'}{\nu}) = 39.8°$$

Squaring and equating the formulas (6.29) we have

$$p'^2_e = \frac{h^2}{c^2}(\nu^2 + \nu'^2) \qquad (6.30)$$

The energy of the electron and its momentum are related by the formula (see Sec. 6.1.4)

$$W_e'^2 = (p'_e c)^2 + (m_0 c)^2$$

that, using (6.30), can be written

$$W_e'^2 = (p_e'c)^2 + (m_0 c)^2 = h^2 (\nu^2 + \nu'^2) + (m_0 c)^2$$

or

$$W_e' = \sqrt{h^2 (\nu^2 + \nu'^2) + (m_0 c)^2}$$

Equating the sum of the energies of the photon and the electron before the collision to its value after the collision, we have

$$h\nu + m_0 c^2 = h\nu' + \gamma m_0 c^2$$

Hence

$$\Delta W_{ph} = h(\nu - \nu') = (\gamma - 1)m_0 c^2$$

Therefore for the velocity of the electron after the collision we can find

$$(\gamma - 1) = 0.034 \qquad \gamma = \frac{1}{\sqrt{1 - \dfrac{v^2}{c^2}}} = 1.034 \qquad v = 0.2544c = 7.6 \cdot 10^7 \frac{m}{\sec}$$

6.2.10 Compton 3

A beam of photons of wavelength $\lambda = 12$ pm is scattered by electrons of rest mass m_0. After collision the photons depart of an angle θ from the initial direction. After recoil the electrons move along a direction making an angle $\varphi = 59.6°$ with the x axis. Find the wavelength λ' and the direction θ of the photons after the collision, and the kinetic energy in eV gained by the electrons (Fig. 6. 18).

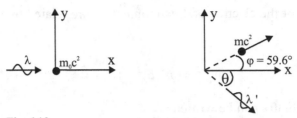

Fig. 6.18

Solution:
We have

$$W_{ph} = \frac{hc}{\lambda} = 103.5 \text{ keV} \qquad (6.31)$$

From principle of conservation of linear momentum the following formulae follow

$$p_{ph} = p_{ph}{}' \cos\theta + p_e \cos\varphi \qquad p_{ph}{}' \sin\theta = p_e \sin\varphi$$

Removing θ by combining the previous formulae, we have

$$p_{ph}{}'^2 = p_e^2 \sin^2\varphi + (p_{ph} - p_e \cos\varphi)^2$$

or

$$p_{ph}{}'^2 - p_{ph}{}^2 = p_e(p_e - 2p_{ph}\cos\varphi)$$

Using relation between linear momentum of a photon with its energy, the previous formula becomes

$$\frac{W_{ph}{}'^2 - W_{ph}{}^2}{c^2} = p_e(p_e - 2\frac{W_{ph}}{c}\cos\varphi) \qquad (6.32)$$

By the energy conservation

$$W_{ph}{}' = W_{ph} - T$$

formula (6.32) becomes

$$\frac{T^2 - 2W_{ph}\,T}{c^2} = p_e\left(p_e - 2\frac{W_{ph}}{c}\cos\varphi\right)$$

$$T^2 - 2W_{ph}\,T = cp_e\,(cp_e - 2W_{ph}\cos\varphi) \tag{6.33}$$

Using the relation between linear momentum and kinetic energy of the electron (see Sec. 6.1.4)

$$p_e c = \sqrt{T(T + 2m_0 c^2)}$$

formula (6.33) becomes

$$T^2 - 2W_{ph}\,T = \sqrt{T(T + 2m_0 c^2)}\left[\sqrt{T(T + 2m_0 c^2)} - 2W_{ph}\cos\varphi\right]$$

that, after some simple but tedious algebraic operations, becomes

$$T = \frac{2W_{ph}{}^2\,m_0 c^2 \cos^2\varphi}{[(m_0 c^2 + W_{ph})^2 - W_{ph}{}^2\cos^2\varphi]} = 7.4 \text{ keV} \tag{6.34}$$

where we have used (6.31) for W_{ph}, the assigned value of φ and the standard constants m_0 and c.
Hence from

$$T = W_{ph} - W_{ph}{}'$$

it follows

$$W_{ph}{}' = W_{ph} - T = 96.1 \text{ kev}$$

and

$$\lambda' = \frac{hc}{W_{ph}{}'} = 12.9 \text{ pm}$$

Finally from standard Compton's formula

$$\lambda' = \lambda + \frac{h}{m_0 c}(1 - \cos\theta) \qquad A = \frac{h}{m_0 c} = 2.4263 \text{ pm}$$

the value of θ can be calculated

$$\frac{\lambda'-\lambda}{A}=1-\cos\theta \qquad \cos\theta=1-B \qquad B=\frac{\lambda'-\lambda}{A}$$

$$\theta=\arccos(1-B)=51.6°$$

6.2.11 Doppler 1

A source of monochromatic light of frequency $\nu = 1.0\cdot10^{14}$ Hz emits a beam directed from A to B where is stationary an observer. The source is moving backward from A with the velocity v_S along the line that includes the segment A-B (Fig. 6.19). Find the value of the v_S which satisfies the relation

$$\Delta\nu = \frac{\nu'_C-\nu'_R}{\nu} \geq 0.01$$

where

$$\nu'_C =\nu(1-\frac{v_S}{c}) \qquad \nu'_R =\nu\frac{1-\dfrac{v_S}{c}}{\sqrt{1-\dfrac{v_S^2}{c^2}}}$$

are the classic and relativistic formulae for the calculus of the frequency measured by the observer.

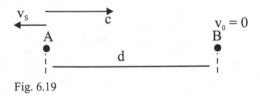

Fig. 6.19

Find the values of ν'_C and ν'_R that satisfy the assigned condition for $\Delta\nu$. Give the plot of ν'_C and ν'_R versus the velocity v_S.

Solution:

We are on the condition of box **A** of Fig. 6.3 (see Sec. 6.1.6b) with v_S

replaced by v. Indeed we are dealing with a relative velocity without regard to who (source or observer) is moving and who is stationary. For the expected classical and relativistic values of the Doppler frequency v'_C and v'_R we write

$$v'_C = v(1-\alpha) \qquad \alpha = \frac{v_S}{c}$$

$$v'_R = v\frac{1-\alpha}{\sqrt{1-\alpha^2}} = v(1-\alpha)(1+\frac{1}{2}\alpha^2) = v(1-\alpha) + v\frac{1}{2}\alpha^2(1-\alpha)$$

Looking for

$$\Delta v = \frac{v'_C - v'_R}{v} = -\frac{1}{2}\alpha^2(1-\alpha) = 0.01$$

we encounter the following equation

$$\alpha^3 - \alpha^2 - 0.02 = 0$$

whose solutions are

$$\alpha = \begin{vmatrix} 0.9791 \\ 0.1537 \\ -0.1329 \end{vmatrix}$$

We are looking for a positive value of α. A negative value would be used if condition **B** of Fig. 6.3 (see Sec. 6.1.6b) was considered.
Using the first value of α we find

$$\alpha_1 = 0.979 \qquad v_1 = 2.94 \cdot 10^8 \text{m/sec}$$

$$v'_C = 0.02 \cdot 10^{14} \text{Hz} \qquad v'_R = 0.10 \cdot 10^{14} \text{Hz} \qquad \Delta v = 0.08$$

and with the second root

$$\alpha_2 = 0.154 \qquad v_1 = 4.61 \cdot 10^7 \text{m/sec}$$

$$v'_C = 0.85 \cdot 10^{14} \text{Hz} \qquad v'_R = 0.86 \cdot 10^{14} \text{Hz} \qquad \Delta v = 0.01$$

Therefore the requested values are those retrieved with the second root (Fig. 6.20).

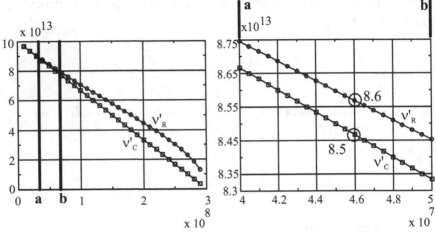

Fig. 6.20 Frequencies (Hz) versus velocity (m/sec)

6.2.12 Doppler 2

In the inertial system *xyz* a beam of light source S of frequency v is incident on the mirror M with an angle $\theta = 30°$ (Fig. 6.21). Both the source S and the mirror M are stationary in the system *xyz*. An observer, stationary in the system *x'y'z'*, that moves with the relative velocity v looking at the beam reflected by the mirror in the *xyz* system, measures an angle of

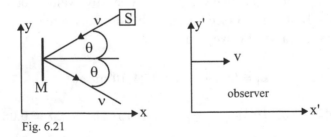

Fig. 6.21

reflection θ' and a frequency v'. Find these values when the two system are moving toward each other and when they are moving away from one another assuming v = 0.5 c. Give the plot of these values varying v/c in the range (0.1, 0.9).

Solution:
The following formulae must be used

$$\nu'=\nu\gamma\left(1-\frac{v}{c}\cos\theta\right) \qquad \cos\theta'=\frac{\cos\theta-\dfrac{v}{c}}{1-\dfrac{v}{c}\cos\theta} \qquad \gamma=\frac{1}{\sqrt{1-\dfrac{v^2}{c^2}}}$$

where the sign of the ratio v/c must be changed if the inertial systems are moving toward each other. The numerical values and the plots requested are in Fig. 6.22, Fig. 6.23 and Fig. 6.24.

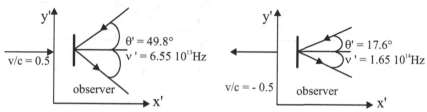

Fig. 6.22

6.2.13 Doppler 3

An observer in the system $x'y'z'$ is looking at a beam of light coming from the system xyz (Fig. 6.25). The system $x'y'z'$ is moving toward the system xyz with a velocity v = 0.8 c. The beam of light appears, in the system $x'y'z'$, to be moving up and down between the angles $\theta'_1 = 0°$ and $\theta'_2 = 36.9°$. The observers measures a greater frequency $\nu'_1 = 3.0\cdot10^{14}$ Hz when the beam has the angular position $\theta'_1 = 0°$ but a lower frequency $\nu'_2 = 1.67\cdot10^{14}$ Hz when the beam has the angular position $\theta'_2 = 36.9°$. Find the corresponding values of θ_1, θ_2, ν_1 and ν_2 in the system xyz.

Solution:
We have to use the following formulae (see Sec. 6.1.6b)

$$\nu=\nu'\gamma\left(1+\frac{v}{c}\cos\theta'\right) \qquad \cos\theta=\frac{\dfrac{v}{c}+\cos\theta'}{1+\dfrac{v}{c}\cos\theta'}$$

with v = - 0.8 c.

For $\theta'_1 = 0°$ and $v'_1 = 3.0 \cdot 10^{14}$ we find $\theta_1 = 0°$ and $v_1 = 1.0 \cdot 10^{14}$ Hz. We are in the condition of the *longitudinal* relativistic Doppler effect.
For $\theta'_2 = 36.9°$ and $v'_2 = 1.67 \cdot 10^{14}$ Hz we find $\theta_2 = 90°$ and $v_2 = 1.0 \cdot 10^{14}$ Hz. We are in the condition of the *transverse* relativistic Doppler effect.

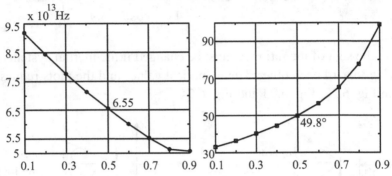

Fig. 6.23 v' and θ' versus v/c when the two inertial systems are receding

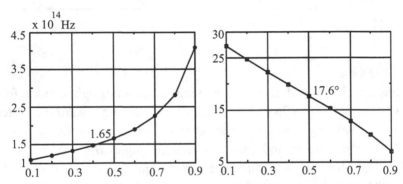

Fig. 6.24 v' and θ' versus v/c when the two inertial systems are approaching

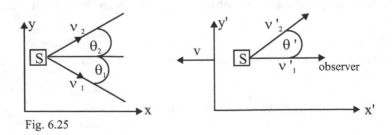

Fig. 6.25

6.2.14 Bradley

The apparent displacement of a star (Gamma Draconis) led Bradley, in 1725, to discover stellar *aberration*: the moving of stars in elliptical paths on the celestial sphere if observed during a whole year. The aberration is greatest for the stars whose line of sight is at right angle with the Earth's orbital speed around the Sun. Bradley realized that for these stars the ellipse had a semi major axis $\Delta\theta = 20.5''$(one sixtieth of a minute), called constant of aberration; equal to v/c, the ratio of the Earth's orbital speed (v = 29770 m/sec) to the speed of light. So Bradley gave an approximate value of the light speed as c = v/ $\Delta\theta$ = 29770/10^{-4} = 2.977·10^8 m/sec. Assuming that the star is emitting a beam of light along the y axis (θ = 90°) in the inertial system xyz, using the relativistic theory of Doppler effect and knowing the values of v and c find the value of $\Delta\theta$ (Fig. 6.26). If the stars emits a monochromatic radiation of frequency v = 5.10·10^{14} Hz, verify the values of the frequencies that would be observed at time t and at time t+6 months in the $x'y'z'$ system.

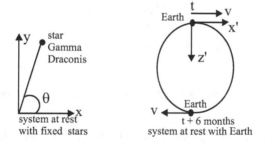

Fig. 6.26 The z axis of the system xyz and the y' axis of the x'y'z' are normal to the plane of figure

Solution:

*First way.*The values of θ' and θ'' in $x'y'z'$ at the time t and at time $t + 6$ months (Fig. 6. 27) are given by the formulae

$$\theta' = \arccos\frac{\cos\theta - \dfrac{v}{c}}{1 - \dfrac{v}{c}\cos\theta} = 90.0057° \qquad \theta'' = \arccos\frac{\cos\theta + \dfrac{v}{c}}{1 + \dfrac{v}{c}\cos\theta} = 89.9943°$$

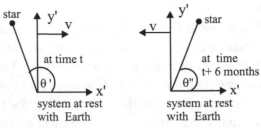

Fig. 6.27

Hence we have (Fig. 6.28)

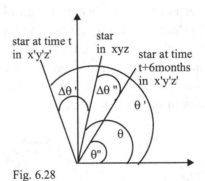

Fig. 6.28

$$\Delta\theta'=\theta'-\theta=20.48''\qquad \Delta\theta''=\theta''-\theta=-20.48''$$

Another way. When the system $x'y'z'$ is moving away from the system xyz with relative velocity v the corresponding angles θ' and θ are related (at time t) by the formula

$$\cos\theta=\frac{\dfrac{v}{c}+\cos\theta'}{1+\dfrac{v}{c}\cos\theta'}\qquad(6.35)$$

with v positive. When the system $x'y'z'$ is moving toward the system xyz with relative velocity v the corresponding angles θ'' and θ are related (at time $t+6$ months) by the formula

$$\cos\theta = \frac{-\frac{v}{c} + \cos\theta'}{1 - \frac{v}{c}\cos\theta'} \qquad (6.36)$$

with the positive value of v.
Formula (6.35) can be written

$$\cos\theta = (\frac{v}{c} + \cos\theta')(1 - \frac{v}{c}\cos\theta') = \frac{v}{c} + \cos\theta' - \frac{v}{c}\cos^2\theta' \qquad (6.37)$$

and (6.36)

$$\cos\theta = (\frac{v}{c} + \cos\theta'')(1 + \frac{v}{c}\cos\theta'') = -\frac{v}{c} + \cos\theta'' + \frac{v}{c}\cos^2\theta'' \qquad (6.38)$$

Remembering (Fig. 6.28) that $\theta = 90°$, θ' and θ'' have values close to $90°$ with $\Delta\theta'$ and $\Delta\theta''$ very small and putting

$$\theta = \theta' - \Delta\theta' \qquad \theta = \theta'' - \Delta\theta''$$

formulae (6.37) and (6.38) become

$$\cos(\theta' - \Delta\theta') = \Delta\theta' = \frac{v}{c} + \cos\theta' - \frac{v}{c}\cos^2\theta' = \frac{v}{c} = \frac{29770}{3\cdot10^8} = 20.48''$$

$$\cos(\theta'' - \Delta\theta'') = \Delta\theta'' = -\frac{v}{c} + \cos\theta'' + \frac{v}{c}\cos^2\theta'' = -\frac{v}{c} = -20.48''$$

where v is the speed at which the Earth moves around the Sun
Apart from the sign an observer (like Bradley) in the $x'y'z'$ system measures 20.48'' for the constant of aberration.
The frequencies observed at time t and at time t+6 months are the same as the frequency emitted from the star because we are in condition of transverse Doppler effect and the orbital speed v is not sufficient to change the value of γ from 1. Clearly (6.25)

$$\nu' = \gamma\nu = \frac{\nu}{\sqrt{1 - \frac{v^2}{c^4}}} = \frac{\nu}{\sqrt{1 - (10^{-4})^2}} = \nu$$

Appendix 1

Fourier Series and Integrals

Let's have a periodic function $f(t)$ with a period T_0 and a frequency ν_0 (Fig. A1.1)

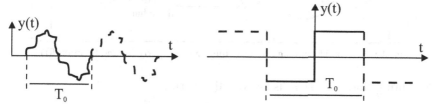

Fig. A1.1 Examples of periodic functions

$$f(t) = f(t+T_0) = f(t+\frac{1}{\nu_0}) \qquad (A1.1)$$

The *Fourier's theorem* (whose proof is given by mathematicians) states that under certain *sufficient* conditions the function $f(t)$ can be expanded into the following trigonometric series of infinite terms

$$f(t) = a_0 + a_1 \cos(2\pi\nu_0 t) + b_1 \sin(2\pi\nu_0 t) +$$

$$+a_2 \cos(4\pi\nu_0 t) + b_2 \sin(4\pi\nu_0 t) + ... + a_n \cos(n2\pi\nu_0 t) + b_n \sin(n2\pi\nu_0 t)$$

The terms of the series are called harmonics because they are simple sinusoidal waves. They have different amplitudes and their frequencies are multiples of the fundamental frequency ν_0 of $f(t)$ (Fig. A1.2).

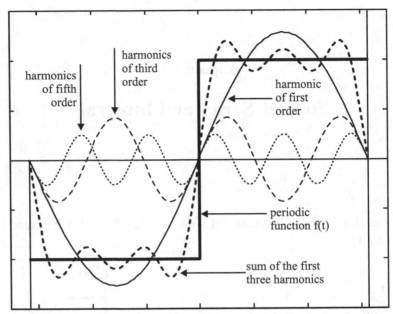

Fig. A1.2 If 1000 harmonics are used their sum properly fits the function f(t)

Assuming that $\nu_0 = 1/T_0$ is known, this series

$$f(t) = \sum_{n=-\infty}^{\infty} \left[a_n \cos(n2\pi\nu_0 t) + b_n \sin(n2\pi\nu_0 t) \right] \qquad \text{(A1.2)}$$

can also be written

$$f(t) = \frac{A_0}{2} + \sum_{n=1}^{\infty} \left[A_n \cos(n2\pi\nu_0 t) + B_n \sin(n2\pi\nu_0 t) \right]$$

with

$$A_n = 2\nu_0 \int_0^{1/\nu_0} f(t)\cos(n2\pi\nu_0 t)dt \qquad B_n = 2\nu_0 \int_0^{1/\nu_0} f(t)\sin(n2\pi\nu_0 t)dt$$

or more formally

$$f(t) = \frac{A_0}{2} + \sum_{n=1}^{\infty} C_n e^{in2\pi\nu_0 t} \qquad C_n = A_n - iB_n \qquad \text{(A1.3)}$$

with

$$C_n = 2\nu_0 \int_0^{1/\nu_0} f(t) e^{-in2\pi\nu_0 t} dt \qquad \text{(A1.4)}$$

If $f(t)$ represented the amplitudes of an electric field associated to a light beam, we would have

$$I_n = C_n C_n{}^* = (A_n - iB_n)(A_n + iB_n) = A_n{}^2 + B_n{}^2$$

$$\rho_n = \sqrt{A_n{}^2 + B_n{}^2}$$

(A1.5)

where I_n an ρ_n would be the intensity and the amplitudes of the harmonic of the order n.

Then the total intensity would be

$$I = \sum_{n=1}^{\infty} I_n$$

(A1.6)

But a beam of light emitted from a usual (thermal, for example) source or from a "quasi-monochromatic" source neither is *really monochromatic* (represented by a single harmonic) nor is composed of a number of discrete multiple frequencies of a fundamental frequency, as is required by the Fourier's theorem. A beam of *natural* light from a usual source contains a continuum of frequencies (remember the colored beam due to the dispersion by a prism) within the range $(4.3, 7.4) \cdot 10^{14}$ Hz. This range for a "quasi-monochromatic" beam is never formed by a single finite value. Therefore the light frequencies must defined as

$$\nu' = \nu + \Delta\nu$$

where $\Delta\nu$ is an infinitesimal quantity and the number of amplitudes $A(\nu)$ tends to infinite as $\Delta\nu$ approaches zero.

The *Fourier's integrals theorem* (whose proof is given by mathematicians) provides the right formulae for this case

$$F(t) = \int_{-\infty}^{\infty} A(\nu)e^{i2\pi\nu t}\,d\nu \qquad A(\nu) = \int_{-\infty}^{\infty} F(t)e^{-i2\pi\nu t}\,dt$$

(A1.7)

These formulae, called Fourier's *transforms*, exhibit some formal likeness to the previous (A1.3) and (A1.4): $f(t)$ was expanded into a sum of infinite discrete values and $F(t)$ is given by an integral that is the limit, as $\Delta\nu$ approaches zero, of the sum of infinite products of amplitudes $A(\nu)$ by

Δv. Their infinite phases are present in the exponent of the base e of the natural system of logarithms. These amplitudes are represented by the function $A(v)$, usually complex, that formally resembles the formula of C_n (A1.4). The product, similar to (A1.5), of $A(v)$ by its complex conjugate

$$I_v = A(v)A^*(v) \tag{A1.8}$$

is called, in the Fourier's jargon, *power spectrum*. This is the *spectral intensity* (see Sec. 6.1.2) used in the Planck's spectral distribution and defined as power per unit surface and unit frequency.

The definitions of the intensity and spectral intensity are well-founded on the content of the Poynting vector (Fig. A1.3)

$$\vec{S} = \frac{1}{\mu_0}\vec{E}x\vec{B} \tag{A1.9}$$

whose direction is equal to that of the velocity of propagation of the radiation. Its magnitude, using only the electric field in the formula, is

$$S = \frac{1}{2}\varepsilon_0 cE^2$$

The quantity ε_0 and μ_0 are the dielectric constant and the permeability of the free space. The quantity S, measured in watt/m^2, is what we call intensity or, using Fourier's theorems, spectral power density.

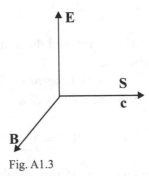

Fig. A1.3

Natural (our eyes, for example) and technological (a photodiode, for example) detectors are unable to see the oscillation of the light wave. Light waves have frequencies of the order of 10^{14} Hz. The corresponding period T is of the order of 10^{-14} sec or 0.01 picosec. Detectors are unable to perceive this extremely small amount of time. What they perceive is the intensity.

The energetic quantity identifying a light source is its power (measured in watt). The energetic quantity that identifies an observation of light on a surface is its intensity (measured in watt/m^2), if we are considering an *abstract* single frequency, or the spectral intensity (measured in watt/(m^{2}Hz)), if we are considering the small or large spectrum of a light beam.

Appendix 2

Superposition of Plane Harmonic Waves

The sum f of following plane harmonic waves

$$f_0 = A \sin \alpha = A e^{i\alpha}$$

$$f_1 = A\varepsilon e^{i(\alpha+\varphi)} \qquad f_2 = A\varepsilon^2 e^{i(\alpha+2\varphi)} \quad ... \quad f_n = A\varepsilon^n e^{i(\alpha+n\varphi)} \qquad \text{(A2.1)}$$

is

$$f = f_0 (1 + \varepsilon e^{i\varphi} + \varepsilon^2 e^{i2\varphi} + ... + \varepsilon^n e^{in\varphi}) \qquad \text{(A2.2)}$$

Multiplying both sides by $\varepsilon^{i\varphi}$ we obtain

$$f\varepsilon e^{i\varphi} = f_0 (\varepsilon e^{i\varphi} + \varepsilon^2 e^{i2\varphi} + \varepsilon^3 e^{i3\varphi} + ... + \varepsilon^{n+1} e^{i(n+1)\varphi}) \qquad \text{(A2.3)}$$

Subtracting (A2.3) from (A2.2) it follows

$$f (1 - \varepsilon e^{i\varphi}) = f_0 (1 - \varepsilon^{n+1} e^{i(n+1)\varphi}) \qquad \text{(A2.4)}$$

$$f = f_0 \frac{1 - \varepsilon^{n+1} e^{i(n+1)\varphi}}{1 - \varepsilon e^{i\varphi}} \qquad \text{(A2.5)}$$

First Hypothesis
If $\varepsilon < 1$ and n is very great, putting $m = n + 1$, (A2.5) becomes

$$f = f_0 \frac{1}{1-\varepsilon\, e^{i\varphi}} \tag{A2.6}$$

and its complex conjugate is

$$f^* = f_0 \frac{1}{1-\varepsilon\, e^{-i\varphi}} \tag{A2.7}$$

From (A2.6) and (A2.7) we have

$$I_r = \frac{f f^*}{f_0^2} = \frac{1}{1-\varepsilon\, e^{i\varphi}} \frac{1}{1-\varepsilon\, e^{-i\varphi}} = \frac{1}{1-\varepsilon\, e^{-i\varphi} - \varepsilon\, e^{i\varphi} + \varepsilon^2} =$$

$$= \frac{1}{1+\varepsilon^2 - 2\varepsilon\cos\varphi} \tag{A2.8}$$

This function has the maximum

$$I_{r\,\text{max}} = \frac{1}{(1-\varepsilon^2)} \tag{A2.9}$$

if

$$\cos\varphi = 1 \qquad 2p\pi \qquad p = 0,1,2,\ldots$$

and the minimum

$$I_{r\,\text{min}} = \frac{1}{(1+\varepsilon^2)} \tag{A2.10}$$

if

$$\cos\varphi = -1 \qquad \varphi = (2p+1) \qquad p = 0,1,2,\ldots$$

Second Hypothesis
If $\varepsilon = 1$ and n is finite, putting $N = n + 1$, (A2.5) becomes

$$f = f_0 \frac{1- e^{iN\varphi}}{1-e^{i\varphi}} \tag{A2.11}$$

and its complex conjugate is

$$f^* = f_0 \frac{1 - e^{-iN\varphi}}{1 - e^{-i\varphi}} \qquad (A2.12)$$

From (A2.11) and (A2.12) we have

$$I_r = \frac{f f^*}{f_o^2} = \frac{1 - e^{iN\varphi}}{1 - e^{i\varphi}} \frac{1 - e^{-iN\varphi}}{1 - e^{-i\varphi}} = \frac{1 - e^{iN\varphi} - e^{-iN\varphi} + 1}{1 - e^{i\varphi} - e^{-i\varphi} + 1} =$$

$$= \frac{1 - \cos N\varphi}{1 - \cos\varphi} = \frac{1 - \cos^2 \frac{N\varphi}{2} + sen^2 \frac{N\varphi}{2}}{1 - \cos^2 \frac{\varphi}{2} + sen^2 \frac{\varphi}{2}} = \frac{sen^2 \frac{N\varphi}{2}}{sen^2 \frac{\varphi}{2}} \qquad (A2.13)$$

The last result (A2.13) gives the relative intensity due to the factor of interference (see Sec. 5.1.2) using a grating of 2, or 3, ..., or N slits.